STP 1041

Asphalt Concrete Mix Design: Development of More Rational Approaches

William Gartner, Jr., Editor

ASTM
1916 Race Street
Philadelphia, PA 19103

Library of Congress Cataloging-in-Publication Data

Asphalt concrete mix design: development of more rational approaches/ William Gartner, Jr., editor.
  (STP: 1041)
  "ASTM publication code number 04-010410-08"—T.p. verso.
  Papers presented at the symposium held Dec. 9, 1987 in Bal Harbour, Fla., and sponsored by ASTM Commitee D-4 on Road and Paving Materials.
  Includes bibliographical references.
  1. Asphalt concrete—Congresses. I. Gartner, William, 1925–
  II. ASTM Committee D-4 on Road and Paving Materials. III. Series: ASTM special technical publication: 1041.
  TE200.A85 1989
  625.8'5—dc20   89-17603
                                                                                 CIP

Copyright © by AMERICAN SOCIETY FOR TESTING AND MATERIALS 1989

NOTE

The Society is not responsible, as a body,
for the statements and opinions
advanced in this publication.

Peer Review Policy

Each paper published in this volume was evaluated by three peer reviewers. The authors addressed all of the reviewers' comments to the satisfaction of both the technical editor(s) and the ASTM Committee on Publications.

The quality of the papers in this publication reflects not only the obvious efforts of the authors and the technical editor(s), but also the work of these peer reviewers. The ASTM Committee on Publications acknowledges with appreciation their dedication and contribution of time and effort on behalf of ASTM.

Printed in Baltimore, MD
October 1989

Foreword

The papers in this publication, *Asphalt Concrete Mix Design: Development of More Rational Approaches,* were presented at the symposium on Development of More Rational Approaches to Asphalt Concrete Mix Design Procedures held 9 December 1987 in Bal Harbour, Florida. The symposium was sponsored by ASTM Committee D-4 on Road and Paving. William Gartner, Jr., The Asphalt Institute, is editor of this publication.

Contents

Overview 1

The Evolution of Asphalt Concrete Mix Design—WILLIAM H. GOETZ 5

Asphalt-Aggregate Mixtures Analysis System: Philosophy of the Concept—
HAROLD L. VON QUINTUS, JIM SCHEROCMAN, AND CHUCK HUGHES 15
Discussion 36

A Comprehensive Asphalt Concrete Mixture Design System—C. L. MONISMITH,
F. N. FINN, AND B. A. VALLERGA 39

The Role of Asphalts in Rational Mix Design and Pavement Performance—
BYRON E. RUTH, MANG TIA, AND KWASI BADU-TWENEBOAH 72

Asphalt Mix Design and the Indirect Test: A New Horizon—GILBERT Y. BALADI
AND RONALD S. HARICHANDRAN 86

Toward Maximum Performance Mix Design for Each Situation—
WILLIAM O. YANDELL AND ROBERT B. SMITH 106

Asphalt Concrete Mix Design in the Caribbean—RAYMOND CHARLES 115

Overview

Asphalt concrete mix design procedures currently in use are empirical in concept and primarily function as a tool to select an asphalt content for an aggregate gradation selected on the basis of local experience. Over the period of time they have been used, strong correlations have been developed between the mix design parameters measured and pavement performance. But, because they are empirically based, they are valid only for those materials, mixtures, and field conditions for which the correlations with performance were established.

In the past 15 to 20 years, major changes have occurred, and are still occurring, in the loading conditions under which pavements are being required to perform. Legal axle loads and tire pressures have increased dramatically, as have the percentages of heavy axle loads on our major highways. Changes in material properties—and therefore mixture properties—have also been occurring. With the development and increased use of asphalt modifiers, the properties of today's and tomorrow's pavement mixtures may bear little resemblance to the properties of mixtures for which empirical correlations have been developed. When conditions change, new correlations must be established. Such correlations are costly and take years to develop and confirm. Thus, the need for a more rational approach is evident.

This ASTM Special Technical Publication has been published as the result of a symposium held in Bal Harbour, Florida, in 1987 organized to present a review of current efforts to develop more rational approaches to asphalt concrete mix design and to encourage further effort in this direction. It is the outgrowth of the work of several subcommittees of ASTM Committee D-4 on Road and Paving Materials. It presents a historical review of mix design procedures, answering the questions: "Where are we?" and "How did we get here?" It also includes papers describing current efforts to develop rational mix design procedures that answer the following questions. "What distress modes need to be considered, and are they independent or interdependent?" "What critical stresses need to be considered, and how can we design a mix to accommodate them?" "Is an integrated mix design system possible or practical?" "To what extent is it necessary or desirable to duplicate field conditions?" (that is, aggregate orientation, asphalt absorption, asphalt hardening, service environment, etc.).

The paper by Professor Goetz reviews the historical development of the empirical methods now in common use, along with the theoretical concepts on which they are based, and points out that while they have served usefully, they lack a fundamental, rational point of view and serve to emphasize the need for a system that relates engineering properties to mix parameters and translates to design that would resist failure in all of its various modes.

The Von Quintus et al. paper presents the philosophy of the AAMAS (Asphalt-Aggregate Mixture Analysis System) concept and discusses those factors that must be included in the design of optimum paving mixtures based on performance-related criteria. The focus of the development of AAMAS was to develop new laboratory mixture design procedures or modify existing procedures based on engineering properties, structural behavior, and pavement performance.

Specific items addressed in the paper include compaction of laboratory mixtures to simulate the characteristics of mixtures placed in the field, preparation and mixing of materials in the laboratory to simulate the asphalt concrete plant production process, simulation of the long-term effects of traffic and the environment (this includes accelerated aging and densification of the mixes caused by traffic), and the conditioning of laboratory samples to simulate the effects of moisture-induced damage and hardening of the asphalt.

The AAMAS concept is applicable to hot-mixed asphalt concrete and includes mixture variables such as binders, aggregates, and fillers used in the construction of asphalt concrete pavements. AAMAS currently excludes such materials or layers as open-graded friction courses and drainage layers.

The paper by Monismith et al. represents an alternative approach to an integrated mix design procedure. Its focus is based on the concept that the design of an asphalt concrete mix consists, essentially, of the following steps:

1. select type and gradation of aggregate;
2. select type and grade of asphalt, with and without a modifier; and
3. select proportionate amount of asphalt in asphalt/aggregate blend.

Monismith incorporates these steps into a general framework for design, which serves as the basis for the mix design procedure presented in this paper.

Essentially, the system consists of a series of subsystems in which the mix components and their relative proportions are selected in a step-by-step procedure to produce a mix that can then be tested and evaluated to ensure that it will perform adequately in the specific pavement section for which it has been formulated. The latter evaluation phase includes the influence of environmental factors, effects of traffic, and the consequence of the anticipated structural cross-section design at the designated site in the following distress modes: fatigue, rutting, thermal cracking, and raveling. This paper also includes a discussion of the important factors associated with the various steps of the design process and recommended test procedures to be followed.

The paper by Ruth et al. focuses more on the role of the asphalt binder and includes a discussion on the selection of asphalts and polymer modifiers, illustrating deficiencies in current test methods and specifications. Because of his concern about brittle cracking at low temperatures, he advises that low-temperature viscosity tests should be incorporated in all asphalt specifications. A comprehensive model asphalt specification is presented that includes viscosity tests at 15 and 25°C. An example is given to demonstrate how the viscosity-temperature relationships enhance our ability to interpret asphalt, polymer modified asphalt, and mix behavior at low temperature.

The remaining papers focus more on modifications to currently used procedures or equipment. In the Baladi and Harichandran paper, the features of a new indirect tension test apparatus is introduced. Analytical models to reduce the test data and to calculate the structural properties of asphalt mixes are presented and discussed. A summary of the findings along with the resulting statistical equations are also presented. It is shown that the structural properties of asphalt mixes obtained from the indirect tension test using the new apparatus are consistent and the test data are reproducible.

Yandell and Smith focus on the specific failure modes of rutting and cracking and suggest looking more closely at the plastic, as well as the elastic behavior of asphalt concrete mixtures. They present data from a field trial site near Sidney, Australia, to demonstrate theoretically how both rutting and fatigue cracking life can be greatly extended by adjusting the elastic and plastic properties of asphalt concrete pavement. Based on simulated temperature changes, they found that maximum rutting and cracking life can be achieved by grading the asphaltic concrete properties from stiff nonplastic at the surface down to soft elasto-plastic at the bottom of the asphalt concrete.

Mr. Charles also focuses on rutting and fatigue cracking and presents a structural analysis of full-depth asphalt pavement sections used to generate strain-stiffness profiles for both types of distress. These profiles were combined with the selected criteria to establish a target mix stiffness. Through trial testing with a selected binder, and use of the Shell Nomograph analysis, it is shown that the target mix stiffness can be converted into stability criteria that can be substituted into Asphalt Institute criteria to provide a complete set of mix design specifications.

The papers in this ASTM Special Technical Publication have clearly answered the question that the symposium was intended to answer. The empirical methods currently being used are not adequate for today's loading conditions nor are they adequate to address the altered binder properties we can expect with the increased use of modifiers. New methods are required that will take into account all modes of pavement distress. These new methods should be based on fundamental properties that will permit extrapolation or recalculation of design parameters to meet changes in loading conditions or alteration of material properties. The papers in this publication describe several approaches that can be taken.

It is hoped these papers will also generate additional interest in the development and implementation of rational design procedures. To gain widespread acceptance, however, any new approaches developed must meet certain objectives that have historically been among the objectives of the currently used empirical methods. The new methods must not only determine the nature and proportions of the mix components and be sensitive to mixture or loading variables, they must also exhibit reasonable precision and accuracy, while being simple enough to be run efficiently in our laboratories. In short, they must be both effective and efficient.

William Gartner, Jr.
Asphalt Institute, Tallahassee, Fl 32308; editor.

William H. Goetz[1]

The Evolution of Asphalt Concrete Mix Design

REFERENCE: Goetz, W. H., "**The Evolution of Asphalt Concrete Mix Design,**" *Asphalt Concrete Mix Design: Development of More Rational Approaches ASTM STP 1041*, W. Gartner, Jr., Ed., American Society for Testing and Materials, Philadelphia, 1989, pp. 5–14.

ABSTRACT: The development of mix design procedures is reviewed historically along with the theoretical concepts on which current procedures are based. Durability considerations are included. An evaluation of the strength tests utilized in design procedures is made and the effect of test variables on the selection of a design asphalt content is presented. The currently used standard procedures, while serving a very useful purpose, are shown to be empirically based and lacking from a fundamental and rational point of view. A mix design system is needed that relates engineering properties to mix parameters and translates to design that prevents failure in all of its various modes.

KEY WORDS: mix design, Hubbard-Field method, Hveem method, Marshall method, Smith triaxial method, gyratory testing machine, design theory, strength testing, asphalt concrete, asphalt mix specifications

The first bituminous mixtures used in the United States were mixtures of sand, gravel, broken stone, ashes, and similar materials using coal tar as the binder. This was prior to 1865 and the use was for sidewalks and cross-walks. Mixtures of this kind were laid as highway pavements in Brooklyn, New York, in 1867 and 1868. In the early 1870s, a number of coal tar pavements were laid in Washington, D. C. However, the first asphalt pavement was a sand mixture laid in 1870 in Newark, New Jersey. Although its exact composition is not known, it probably was patterned after natural rock asphalt pavements developed in Paris. Similar pavements were then laid in New York City, Philadelphia, and Washington, D.C. As a result, the use of sheet-asphalt pavements for streets and pavements developed rapidly [1].

While these early sheet-asphalt mixtures were not designed in any real sense, their evolution and efforts to improve their performance in service resulted in the development of certain principles and concepts, particularly with respect to the properties of materials of which they were composed and their function in the mixture. Thus, it came to be recognized that the sand employed should be hard, clean, adequately sharp, and have a suitable gradation. Dense packing of the sand grains and sharpness that provided a keying action was seen as necessary to provide resistance to shoving under traffic. The fact that dust coatings could deter or prevent adhesion of asphalt to the surface of the sand grain and that such adhesion was necessary, particularly in the presence of water, was recognized early. In this respect, the best sands were considered to be those with rough, pitted surfaces, and those with smooth, polished surfaces were to be avoided. The principles that govern voids in compacted aggregates as affected by grain shape and gradation were developed and desirable void characteristics were a subject of evaluation [1].

[1] Professor emeritus, Purdue University, W. Lafayette, IN 47907.

It was recognized early that in order to achieve compactness and denseness, the interstices between the sand grains needed to be filled. Thus, the term "filler" came to be used along with the recognition that fineness was an essential requirement of such a material. Since the sands used contained very little material passing the 200-mesh sieve, specifications for filler required that the bulk of it pass the 200-mesh sieve. Many fine mineral materials were tried, but the superiority of ground limestone and Portland cements surfaced early on the basis of mixture properties and performance. It was rationalized that these materials were superior because they "packed well" when dry and had strong adhesion for asphalt. Ground silica and quartz, used early, fell into disuse because of inferior resistance to water action and the resulting inferior resistance to abrasion by traffic. From a functional point of view, the void-filling property of filler was recognized and employed, but the asphalt-modifying function was not, even though mixtures made with different fillers were seen to have different stiffness and workability characteristics. When binder courses came to be used as intermediate courses between the foundation and the wearing course of sheet-asphalt, larger aggregates were utilized but similar principles prevailed in the formulation of mixtures with some reduction in the quality of aggregates required and in the need or even desirability of small voids. Bituminous concretes, many of them patented mixtures for use as surfaces for roads and streets, were also an early development. Specific gradings were specified to achieve closeness of packing and control of voids in the mineral aggregate.

The asphalt cement used in pavements was a refined one from the beginning. In the early days these refined materials were hard in consistency and required the use of a flux to soften them to form a suitable road binder. Very soon, however, these materials were produced by direct refining to the proper consistency. What is proper consistency and what are the desirable properties of a paving asphalt were delineated from a qualitative point of view. Resistance to change when exposed to climate and traffic, resistance to change when exposed to temperatures required for mixing with aggregates, change in consistency with change in temperature, cementing value, ductility, and purity were recognized properties. The relationships of the properties to service requirements were also delineated to a surprising degree albeit not in a quantified way [1].

Thus, it is apparent that in the evolution of asphalt pavements, certain concepts, principles, and relationships were developed very early that are still regarded as fundamental today. A mixture must be stable in the sense that it does not rut or shove under traffic. Primarily, this stability comes from the aggregate and is related to its size, size distribution, shape, and surface texture. Asphalt may contribute to stability in its function as a cementing medium, but reduces stability if it is present in amounts greater than that required simply for the cementing function. The function of asphalt is primarily to provide durability to the mixture. This is the property that provides resistance to the forces of nature. In the very early literature, statements of the axioms of mixture design suggest that as much asphalt as possible should be used without sacrificing needed stability. The consistency of the asphalt cement must be such that it will not be too brittle in winter or too soft in summer. From a design point of view, a most important factor is determination of the design asphalt content and this requires a compromise between stability and durability requirements [2,3].

Design Asphalt Content Theory

From a theoretical point of view, two basic concepts have been recognized in establishing a design asphalt content for a given aggregate and aggregate grading chosen for use in an asphalt mixture. A void concept was used early in mixture development that theorized that the amount of asphalt should be that just sufficient to fill the voids in the compacted aggregate. This saturation concept, while theoretically sound from a durability point of view, was soon shown to be unsatisfactory because of mixtures that became unstable and slippery in service under traffic [2,3].

That saturation was not a satisfactory concept did not invalidate the theory that the amount of asphalt required is a function of available space in the compacted aggregate structure. Using this approach, the design asphalt content is that which fills the voids to a degree that leaves some room for increase in the asphalt volume due to expansion at summer temperatures and for a decrease in space available as the aggregate densifies with time under traffic. This establishes a lower limit for voids with the upper limit being established on the basis of durability considerations. However, void theory alone cannot be used to establish design asphalt content without some means of determining voids in the compacted mineral aggregate. These voids are a function not only of aggregate characteristics and compactive effort used to achieve density, but also of the nature of the asphalt (lubricating property) and the asphalt content itself. Void theory is utilized as a part of current design procedures primarily from the durability point of view. In applying the concept of void theory, it cannot be assumed that all of the asphalt used is available for filling voids in the aggregate framework. Some of it may be absorbed by the aggregate pores. If absorption is a factor, it may be time-dependent further complicating the determination of how much of the asphalt is available to fill voids.

The second basic concept for the determination of design asphalt content is the surface area theory. This theory is based on the concept that the design asphalt content is that which coats all of the aggregate surface area with an optimum film thickness of asphalt. Based on the fact that aggregate size is a primary factor in the relationship between a given weight of aggregate and its surface area, the surface area concept of design was first employed by the use of empirical formulae based on aggregate grading. The first of these was developed from experience with mixtures of aggregate and road oil in the western United States [2,3].

An early formula used by the California Highway Department was

$$P = 0.015a + 0.03b + 0.17c$$

where

P = percentage of bitumen in the mix by weight,
a = percentage of aggregate retained on No. 10 sieve,
b = percentage of aggregate passing No. 10 sieve and retained on No. 200 sieve, and
c = percentage of aggregate passing No. 200 sieve.

The numerical factors used with a, b, and c are related to the surface area of the aggregate sizes.

The validity of this formula was limited to the local materials in California around which it was developed and by the fact that it employed only two sieve sizes. Other formulae were developed by other states that not only used more sieves for a more accurate expression of surface area, but also recognized other factors needed to make the formula more generally applicable. An example is the Nebraska formula that follows [3].

$$P = AG(0.02a) + 0.06b + 0.10c + Sd$$

where

P = percentage of bitumen residue by weight of mixture at time of laying,
A = absorption factor for aggregate retained on No. 50 sieve,

G = specific gravity correction factor of aggregate retained on No. 50 sieve = 1 for aggregate with apparent specific gravity of 2.62.
a = percentage of aggregate retained on No. 50 sieve,
b = percentage of aggregate passing No. 50 sieve and retained on No. 100 sieve,
c = percentage of aggregate passing No. 100 sieve and retained on No. 200 sieve,
d = percentage of aggregate passing No. 200 sieve, and
S = factor characterizing material passing No. 200 sieve.

In applying the surface area concept, it needs to be recognized that factors other than grading or sieve size determine area to be covered. These include aggregate shape, surface texture, and adsorption. All of these factors, including grading, shape, surface texture, and adsorption, taken together determine the surface capacity of the aggregate that is probably a better term to be used in the concept than surface area. Also, in applying the concept, a complicating factor arises when the very fine part of the aggregate is so fine that it acts to modify asphalt properties and no longer can be conceived as particles to be coated with an asphalt film [3].

The California Division of Highways has carried the surface area concept to a practical conclusion inasmuch as it forms the basis for their current method of design. In extending their original empirical formula, they first increased the accuracy of the surface area determination by increasing the number of sieves used. At the same time, differences in the surface characteristics of aggregate were recognized. An "oil-line" chart was developed that expressed the fact that smooth hard particles have less surface capacity than rough ones, and that smaller aggregate particles must be coated with thinner asphalt films than larger aggregate sizes. While a great improvement over an empirical formula, the procedure had shortcomings in that some prior knowledge of the aggregate was needed in order to select the appropriate oil line and the absorption factor was not directly accounted for. Also, the asphalt viscosity factor was not included.

In recognition of these shortcomings, methods of test were developed to measure directly the surface and absorptive capacity of the aggregates involved. The test applied to the fine aggregate is the Centrifuge Kerosene Equivalent (CKE) test that determines the amount of kerosene retained on the aggregate. From this, a surface constant for the fine aggregate is determined. The surface constant for the coarse aggregate is determined from the amount of Society of Automotive Engineers (SAE) No. 10 oil retained on the aggregate. These two constants are used to determine a constant for the total aggregate that is used with surface area to calculate an oil ratio in pounds of oil per 100 lb of aggregate. This number is then corrected for viscosity of the asphalt in recognition of the fact that higher viscosity or lower penetration asphalts have less lubricating value and a greater amount may be used. It was found from experience that, if the coarse and fine aggregate are similar in surface characteristics, the surface constant for the mix is not significantly different from that of the fine aggregate upon which the CKE test is performed. In this case, only the CKE value and the percentage of the aggregate passing the No. 4 sieve need be determined [3,4].

Mix Design Methods

Mix design methods evolved from the concepts that mixtures must be stable and durable. A strength test was necessary as a measure of stability, and criteria were established by correlation of results obtained on laboratory specimens with the performance of the same mixture under traffic. Duplicating the composition of the field mixture in the laboratory does present some problems, but they are minor compared to producing in the laboratory a specimen of the mixture that truly represents the mixture as it exists in the road.

Methods used in an attempt to do this vary from direct compression, with or without rodding, to hand tamping, impact hammer, kneading action, gyratory shear, vibration, and simulated rolling. The size of the specimen must be determined, not only with respect to maximum aggregate size, but also with respect to the rationality of the strength test.

Strength tests employed in standard mix design methods have varied from punching shear to direct compression without lateral confinement, confined or triaxial, simulated triaxial, and circumferential. Testing variables include temperature and testing speed or rate of deformation. The standard methods apply, almost exclusively, to dense-graded mixtures, and in some cases in a strict sense the criteria are valid only for mixtures falling within specific gradation limits [3,4].

Hubbard-Field Method

One of the earliest methods of mix design was developed for sheet asphalt in the middle 1920s. It employed a punching shear type of test on specimens formed by a combination of hand tamping and direct compression. The maximum load developed under conditions of this test was reported as a stability value. The punching shear test was realistic for the traffic condition of the time, which consisted of wagons with steel wheels. Since 60°C (140°F) was considered to be the most severe temperature to which a paving mixture would normally be subjected in service, the specimens were tested at this temperature. A fast loading rate was used. Bulk density of specimens was determined prior to performing the strength test. When traffic loads increased and mixtures with larger aggregate sizes were used, the test was adapted to these mixtures by increasing specimen size and correspondingly the size of testing molds.

From the test data, plots were made of bulk density versus asphalt content, stability versus asphalt content, percent voids in the total mix versus asphalt content, and percent voids in the aggregate mass versus asphalt content. These plots were used to select a design asphalt content by comparing them against design criteria. The criteria, which included limits on stability values and percent voids in the total mix, were established by comparing test values with pavement performance. The direct procedure was to select the asphalt content at the mid-point of the voids criteria. If this asphalt content gave a mixture that fell within the stability criteria, it was chosen as the design asphalt content. The plot of asphalt content versus voids in the aggregate mass was used to adjust gradation when this was necessary to meet voids criteria.

The Hubbard-Field method filled an early need for a rational design procedure for asphalt-mixtures. As applied to sheet asphalt (generally, mixtures employing asphalt cements and aggregates with at least 65% passing the No. 10 sieve and all passing the No. 4 sieve), it was widely used and its validity was reinforced by the collection of data gained from this wide experience. As applied to coarser mixtures, the shortcomings of the punching shear (shear ring) type of test became apparent. The principles of a design procedure incorporating stability and durability criteria in a systematic quantitative way became well established [3,4].

California or Hveem Method

As stated in the discussion on theory of design asphalt content, the California method is one based on surface area or surface capacity concepts. It is used primarily for dense-graded mixtures. Using the CKE test value and the percent of aggregate passing the No. 4 sieve, or a surface capacity factor determined by a combination of CKE results and a surface capacity test for the coarse aggregate and aggregate surface area, the percent oil to be used is determined by direct calculation using a nomographic chart. This percent oil or oil ratio is corrected for viscosity, again by means of a nomographic chart.

Having established the asphalt content by direct calculation using surface capacity principles, it is then determined if the resulting mixture is satisfactory by testing it for stability, cohesion, void content, and water susceptibility and applying established criteria. Test specimens are formed using a kneading compaction procedure developed to produce realistic specimens as compared to the mixture in service after traffic compaction. Specimens are normally made at the calculated asphalt content and at contents above and below this value. Stability is measured in the Hveem Stabilometer, which is a special type of closed triaxial cell. The test is conducted at 60°C (140°F) at a slow rate of loading. The lateral pressure transmitted to the surrounding fluid during deformation of the specimen is determined. A relative stability value is calculated based on a rigid solid having a value of 100 and that of a liquid having a value of zero. The stabilometer test is not continued to failure and the same specimen is used to determine a cohesiometer value by a test that applies direct tension in bending at 60°C (140°F). Voids determinations are made. A swell test is performed on companion specimens to determine resistance of the compacted mixtures to water. Test results are compared to criteria for stabilometer value, cohesiometer value, and swell established by California for satisfactory mixtures. Voids are not a direct part of mix design, but it is recommended that voids in the total mix be approximately 4%. If criteria are not met, gradation adjustments are made [3,4].

U.S. Army Corps of Engineers or Marshall Method

The mix design method in most general use today was developed by the U.S. Army Corps of Engineers during World War II in answer to the need for a method that could be used in the field for design and control of mixtures for airfield pavements in theaters of operation. To meet the criteria of simplicity and portability, they chose a testing device conceived by Bruce Marshall of the Mississippi Highway Department and a hand method of compaction employing a hammer with a falling weight. Large-scale field tests were conducted using simulated traffic to determine the performance of mixtures of varying composition and to establish the compactive effort needed to fabricate specimens meeting field density.

The Marshall testing apparatus applies a compressive load on the circumference of a cylindrical specimen through semi-circular test heads. The temperature of test is 60°C (140°F), and the testing speed is rapid with the test being continued to failure. Maximum load is recorded as well as the deformation undergone by the specimen in reaching maximum load. These values are designated as Marshall stability and Marshall flow. Density measurements are made before the specimens are tested for strength. Specimens with a range of asphalt content are prepared and tested, usually at the expected optimum asphalt content, and at two values above and two values below this amount. One-half percent increments are normally used.

From the density and voids analysis and the stability and flow tests, the Corps of Engineers plotted bulk density versus asphalt content, stability versus asphalt content, flow versus asphalt content, percent air voids in the total mixture versus asphalt content, and percent voids in the aggregate mass filled with asphalt versus asphalt content. To establish the optimum asphalt content, the asphalt content giving maximum density, that producing maximum stability and those producing values of voids and voids filled at the median value of the criteria range are determined from the plots and averaged. If this asphalt content produces a mixture that meets the established criteria, it is the optimum or design asphalt content. If the mixture does not meet criteria, it is redesigned [3,4].

In applying this design method to highway pavements by the various state highway agencies, the basic compaction and test procedures and design concepts have remained intact. Evolution over the years, during which the method has been applied to a wide range of mixtures and traffic conditions with the consequent enlargement of the data base, has produced some changes. Mixing and compaction temperatures have been related to asphalt viscosity rather than being fixed numbers. The severity of traffic has been categorized into light, medium, and heavy, and the compactive effort required for realistic test specimens in each category has been determined. Design criteria have been established for each traffic category for both surface and base mixtures. A fundamental change from the Corps of Engineers original method has been the dropping of the criteria for percent voids in the aggregate mass filled with asphalt, and the substitution of criteria for percent voids in the mineral aggregate. This was done because it was determined that minimum voids in the aggregate is a more fundamental factor than voids filled. Aggregate gradations that are too dense (low percent voids in the compacted aggregate) do not provide enough room for asphalt [4].

Smith or Asphalt Institute Triaxial Method

The Hubbard-Field, Hveem, and Marshall methods of mix design are all empirical ones with criteria developed from the correlation of laboratory and field results. In recognition of the desirability of a more rational approach, V. R. Smith and the Asphalt Institute applied the principles of the triaxial compression test to the testing of bituminous mixtures [3]. Smith used a closed-system triaxial test wherein the vertical load was applied in static increments and the lateral transmitted pressure was measured when equilibrium had essentially been reached. Specimens of rational size (at least twice as high as their diameter) were formed by spading and double-plunger compaction or by use of the kneading compactor. The test was conducted at room temperature, because under the test conditions the effects of temperature and rate of loading are essentially nonexistent. The data obtained were used to generate a plot of vertical applied load versus lateral transmitted pressure from which the parameters (ϕ), angle of internal friction, and cohesion (c), were calculated. From theoretical considerations, Smith developed the relationship between angle of internal friction and cohesion for varying applied stresses and presented this in the form of supporting power curves.

By testing many mixtures in the laboratory whose performance in the field was known, an evaluation chart was devised that delineated regions of satisfactory and unsatisfactory mixtures in a plot of angle of internal friction versus cohesion. One boundary of the satisfactory region was determined from supporting power considerations that, in turn, is related to the severity of traffic. Thus, the region of satisfactory mixes is increased when traffic conditions are less severe. A second boundary between satisfactory and unsatisfactory mixes derives from the fact that experience showed that a minimum angle of internal friction is required for satisfactory performance. Voids criteria were applied [3].

While the triaxial test is a rational one that generates the measurement of parameters of a more fundamental nature than those measured previously and does use theoretical concepts in applying these measurements to mixture design, in the final analysis, it depends upon correlation between field and laboratory results for application. The measured parameters can be determined in a rational way for known stress conditions in the laboratory, but the pavement is not loaded in a rational way in service and stress conditions in the pavement are unknown.

In spite of such shortcomings from a design point of view, a design based on fundamental measurements was considered progressive in anticipation of the day when stress conditions in the pavement could be more accurately defined and rational strength test data applied. To this end, attempts were made to simplify the measurements of c and ϕ in a triaxial test. The open system of triaxial testing can be used that does not require the sophisticated cell used by Smith [5]. The Bureau of Public Roads, predecessor to the Federal Highway Administration, designed a cell to simplify the test [2]. Purdue University used a vacuum technique that applies confining pressure to the specimen by reducing internal pressure rather than applying pressures higher than atmospheric externally [6]. Using this technique, the confining pressure is limited to a value not exceeding atmospheric pressure. In spite of these efforts, the triaxial test has not been widely used for design purposes, but has been used in the laboratory as a research tool for the most part.

Use of the Gyratory Testing Machine

When the U.S. Army Corps of Engineers in the development of their design procedure and the application of it to increasing loads and tire pressures reached a point where it was not possible or practical to achieve field density in laboratory specimens by impact compaction, they decided to motorize the Texas gyratory device. This resulted in the development of the U.S. Army Gyratory Testing Machine. A test for compaction and shear properties of bituminous mixtures by means of this machine has been standardized as ASTM Test Method for Compaction and Shear Properties of Bituminous Mixtures by Means of the U.S. Corps of Engineers Gyratory Testing Machine (GTM) (D 3387-83).

ASTM D 3387-83 tests for strength properties and compaction and strain indices in a wide range of mixtures, both hot and cold. Test procedures provide for the calculation of compaction properties including unit mass total mix, unit mass aggregate only, and gyratory compactibility index. Also, the test provides for the calculation of shear properties including gyratory stability index, gyratory shear, and gyratory shear factor. Specimens with a range of bitumen contents may be tested and the calculated compaction and shear properties plotted against bitumen content. From these plots, an optimum may be selected.

The Corps of Engineers does utilize this machine for the design of some of their heavy-duty airfield pavements. Kallas of the Asphalt Institute developed gyratory testing machine procedures for selecting design asphalt content of paving mixtures [7]. Kumar and Goetz [8] conducted research on the gyratory testing machine as a design tool and as an instrument for bituminous mixture evaluation. Some state agencies have applied gyratory procedures for design, but to date no method has been standardized and criteria for satisfactory performance under various traffic conditions have not been determined.

Mix Design and Strength Considerations

It is apparent that all of the methods of mix design, including the Hveem and Marshall procedures in current use, are used principally to select a design asphalt content for an aggregate and aggregate grading selected on the basis of experience or specification criteria so derived. All are empirically based. None provide for the direct measurement of durability (except perhaps for the effect of water), but rely on voids considerations, that is, voids total mix and voids in the compacted aggregate, to include the durability factor. There have been problems with the calculation of voids, hinging on how the aggregate absorption factor should be treated. Bulk, apparent, and asphalt-impregnated specific gravity have been used. This matter was largely resolved by Rice in his development of the test for "effective" specific gravity now standardized as ASTM Test Method for Theoretical Maximum Specific Gravity of Bituminous Paving Mixtures (D 2041-78). All of the methods include some type of strength or stability test. In comparing these design procedures, some assessment of the strength test used is pertinent.

Of the tests that have been used as a measure of strength or stability in mix design procedures, the triaxial shear strength test has been recognized as the most fundamental in that it uses field-simulated confined loading where stress conditions are known and statically determinant methods of analysis can be applied [9]. Fundamental shear strength parameters, angle of internal friction, and cohesion are determined.

In the application of triaxial testing to bituminous mixtures, several test and mixture variables have been evaluated. Testing speed within the range that has been used in design procedures has been shown to have little effect on angle of internal friction, but greatly affects cohesion with cohesion increasing with increased testing speed. At constant confining pressure, testing speed has little effect on the compressive strength of mixtures at high temperatures, but strength increases with testing speed as the temperature is lowered. The angle of internal friction is affected minimally by test temperature, but cohesion is greatly reduced as temperature is increased [6].

With respect to mix variables, angle of internal friction decreases as asphalt content increases. Cohesion increases to a maximum value and then decreases as asphalt content is increased. For the unconfined condition, maximum stability (compressive strength) occurs at or near the asphalt content providing maximum cohesion. As the degree of confinement increases, the asphalt content at which maximum stability occurs decreases. Viscosity or penetration of the asphalt has a minor effect on angle of internal friction, but greatly affects cohesion, cohesion decreasing as asphalt viscosity decreases or penetration increases [6].

The open system triaxial test has also been used as a research tool to understand better the way in which bituminous mixtures function as load-carrying materials. Such testing has been used to evaluate strength and volume change characteristics [9]. High confining pressures have been used in an attempt to evaluate the effect of confining pressure on the asphalt content giving maximum strength in open-graded and one-sized mixtures [10]. The test has also been used to evaluate the effect of aggregate shape on the stability of bituminous mixtures [11].

Since the Hveem stabilometer test is a form of the triaxial test, it follows that it should provide a realistic measure of stability. However, the size of the specimen and the fact that air is introduced into the system make it impossible to analyze the results by theoretical methods. A measured or calibrated amount of air is used inside the system, but the air trapped between the rubber membrane and the specimen surface varies with the smoothness of the surface and therefore with mix gradation. A displacement correction made at the end of the test is included in an attempt to account for the effect of this trapped air outside the cell proper. Hannan and Goetz [12] demonstrated that this trapped air acted in the same way as the internal air and therefore open-type mixtures could be tested if the surface voids of the specimen were filled with a non-cohesive mortar.

The Marshall stability test has been shown to be a type of compression test in which there is a degree of confinement induced by friction between the test heads and the specimen surface in contact with them. It also appears that there is no fundamental difference between the circumferential loading used in the Marshall test and the axial loading used in a direct compression test. Comparable strength values will be obtained in the two types of test if the height-diameter ratio and the diameter-thickness ratio are adjusted so that equal degrees of confinement are induced in the two tests. A high degree of correlation has been shown to exist between Marshall flow (deformation at maximum load) and angle of internal direction for a wide range of asphalt mixtures. Since this is the case and the test is a confined one, it has been concluded that the criteria selected by the Corps of Engineers do relate to the fundamental strength properties of a bituminous mixture [13,14].

Closure

Mix design procedures in use today are based on concepts developed very early in the use of bituminous mixtures. Their evolution has progressed to the point that the test methods have been standardized and criteria based on them have reached world-wide acceptance. However, useful as these procedures are, they are empirically based and function mainly to select an asphalt content for an aggregate and aggregate gradation meeting criteria selected on the basis of experience. Being empirically based, strictly speaking, they are valid only for those mixture types and field conditions for which correlation has been established. If conditions change, such as the introduction of different aggregate or asphaltic materials or higher tire pressures are used, correlation must be established for these new conditions.

Thus, the need for a more rational approach is clearly evident. Over the past two decades much progress has been made in measuring fundamental mix properties and in correlating them with mix parameters. The measurement of fundamental mix properties, such as modulus, tensile strength, visco-elastic characteristics, fatigue, and creep can be cited. The progress made in this area has only served to emphasize the need for a more rational mix design procedure or process with focus on the distress modes associated with asphalt pavement and their relationship to mix parameters. With the advancement of mechanistic methods for the structural design of asphalt pavements, the need for a fundamentally based asphalt aggregate mix analysis system related to the engineering properties of the materials involved has become imperative.

References

[1] Smith, F. P. in *American Highway Engineers Handbook,* 1st ed., A. H. Blanchard, Ed., Wiley, New York, 1919, pp. 939–1019.
[2] Vallerga, B. A., "Notes on the Design Preparation and Performance of Asphaltic Pavements," Institute of Transportation and Traffic Engineering, University of California, Berkeley, 1953.
[3] Goetz, W. H. and Wood, L. E. in *Highway Engineering Handbook,* 1st ed., K. B. Woods, Ed., McGraw-Hill, New York, 1960, pp. 18–55 to 18–92.
[4] *Mix Design Methods for Asphalt Concrete,* Manual Series No. 2, The Asphalt Institute, March 1979.
[5] Goetz, W. H. and Schaub, J. H. in *Triaxial Testing of Bituminous-Aggregate Mixtures, ASTM STP 252,* American Society for Testing Materials, Philadelphia, 1959, pp. 51–63.
[6] Goetz, W. H. and Chen, C. C., "Vacuum Triaxial Technique Applied to Bituminous-Aggregate Mixtures," *Proceedings,* Association of Asphalt Paving Technologists, Vol. 19, 1950, pp. 55–81.
[7] Kallas, B. F., "Gyratory Testing Machine Procedures for Selecting the Design Asphalt Content of Paving Mixtures," *Proceedings,* Association of Asphalt Paving Technologists, Vol. 33, 1964, pp. 341–362.
[8] Kumar, A. and Goetz, W. H., "The Gyratory Testing Machine as a Design Tool and as an Instrument for Bituminous Mixture Evaluation," *Proceedings,* Association of Asphalt Paving Technologists, Vol. 43, 1974, pp. 350–371.
[9] Schaub, J. H. and Goetz, W. H., "Strength and Volume Change Characteristics of Bituminous Mixtures," *Proceedings,* Highway Research Board, Vol. 40, 1961, pp. 371–405.
[10] Oppenlander, J. C. and Goetz, W. H., "Triaxial Testing of Bituminous Mixtures at High Confining Pressures," *Proceedings,* Highway Research Board, Vol. 37, 1958, pp. 201–218.
[11] Herrin, M. and Goetz, W. H., "Effect of Aggregate Shape on Stability of Bituminous Mixes," *Proceedings,* Highway Research Board, Vol. 33, 1954, pp. 293–308.
[12] Hannan, R. A. and Goetz, W. H., "Testing Open-Graded Bituminous mixtures in the Hveem Stabilometer," *Proceedings,* 45th Annual Road School, Purdue University, Extension Series No. 100, Vol. 44, No. 2, 1960, pp. 84–99.
[13] Goetz, W. H., "Comparison of Triaxial and Marshall Test Results," *Proceedings,* Association of Asphalt Paving Technologists, Vol. 20, 1951, pp. 200–245.
[14] McLaughlin, J. F. and Goetz, W. H., "Comparison of Unconfined and Marshall Test Results," *Proceedings,* Association of Asphalt Paving Technologists, Vol. 21, 1952, pp. 203–236.

Harold L. Von Quintus,[1] *Jim Scherocman,*[2] *and Chuck Hughes*[3]

Asphalt-Aggregate Mixtures Analysis System: Philosophy of the Concept

REFERENCE: Von Quintus, H. L., Scherocman, J., and Hughes, C., "**Asphalt-Aggregate Mixtures Analysis System: Philosophy of the Concept,**" *Asphalt Concrete Mix Design: Development of More Rational Approaches, ASTM STP 1041*, W. Gartner, Jr., Ed., American Society for Testing and Materials, Philadelphia, 1989, pp. 15–38.

ABSTRACT: This paper presents the philosophy of the asphalt-aggregate mixture analysis system (AAMAS) concept and provides a brief discussion on those factors that must be included in the design of optimum paving mixtures based on performance-related criteria. Typically, the structural design of asphalt concrete pavements is based on assumed material properties (layer stiffness coefficients, resilient modulus, and fatigue or permanent deformation constants or both). After the structural design has been completed, materials are submitted and a mixture design is completed. The question then becomes: Does the as-placed material meet the assumptions used in structural design?
  Certainly, asphalt concrete mixture design and analyses need to be related to those factors that affect pavement performance. The AAMAS objective is to predict in-place properties of the asphalt concrete in the laboratory during the mixture design stage to evaluate mixture behavior and performance under traffic and environmental loads. Thus, the focus for the development of AAMAS was to develop new laboratory mixture design procedures or modify existing procedures based on engineering properties, structural behavior and pavement performance, rather than on empirical numbers such as Marshall or Hveem stability.
  Specific items addressed in the paper include compaction of laboratory mixtures to simulate the characteristics of mixtures placed in the field, preparation and mixing of materials in the laboratory to simulate the asphalt concrete plant production process, simulation of the long-term effects of traffic and the environment (this includes accelerated aging and densification of the mixes caused by traffic), and the conditioning of laboratory samples to simulate the effects of moisture-induced damage and hardening of the asphalt.
  The AAMAS concept is applicable to hot-mixed asphalt concrete and includes mixture variables such as binders, aggregates, and fillers used in the construction of asphalt concrete pavements. The AAMAS currently excludes such materials or layers as open-graded friction courses and drainage layers.

KEY WORDS: mix design, asphalt concrete, asphalt specifications, paving mixtures, performance-related specifications

  The Strategic Highway Research Program (SHRP) plans to develop improved asphalt or new binders or both and performance-related specifications for asphalt concrete materials. Improved procedures and analysis systems are needed for the testing and analyses of asphalt concrete mixtures using these improved or new binders or both to obtain the necessary information to develop performance-related specifications. Unfortunately, current

---
[1] President, Brent Rauhut Engineering, Inc., 8240 Mopac Suite 220, Austin, TX 78759.
[2] Consulting engineer, 11205 Brookridge Drive, Cincinnati, OH 45249.
[3] Virginia Transportation Research Council, P.O. Box 3817, University Station, Charlottesville, VA 22903.

mixture design procedures use different criteria and material properties than those required by most pavement structural performance or response models or both. Therefore, there currently exists a need, as recognized by many highway agencies, to bring structural performance of pavements into the optimization of mixture design, as indicated by Fig. 1.

This paper discusses the philosophy of an asphalt-aggregate mixture analysis system (AAMAS) for evaluating and designing mixtures based on performance-related criteria. Certainly, asphalt concrete mixture design and analyses need to be related to those factors that affect pavement performance. The AAMAS objective is to predict in-place properties of the asphalt concrete in the laboratory during the mixture design stage to evaluate mixture behavior and performance under traffic and environmental loads (Fig. 2).

The AAMAS is currently under development and being funded through the National Cooperative Highway Research Program, NCHRP Project 9-6(1). The focus for the development of AAMAS is to develop new laboratory mixture design procedures or modify existing procedures based on engineering properties, structural behavior, and pavement performance, rather than on empirical numbers such as Marshall and Hveem stabilities.

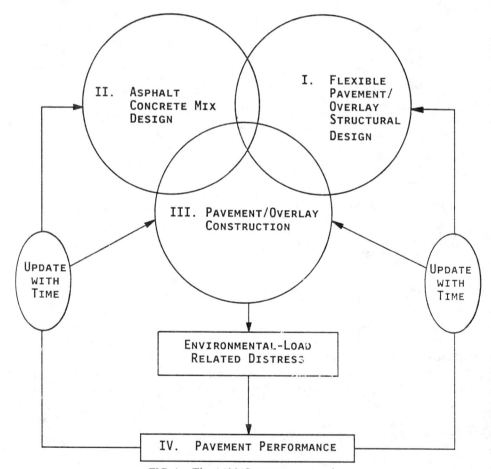

FIG. 1—*The AAMAS concept or triangle.*

VON QUINTUS ET AL. ON ASPHALT-AGGREGATE MIXTURES ANALYSIS SYSTEM 17

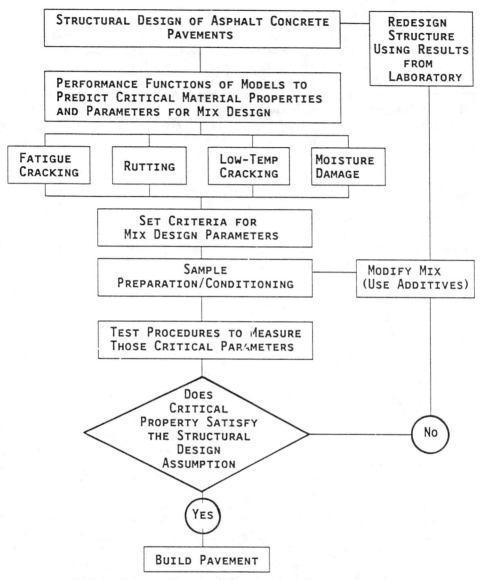

FIG. 2—*Conceptual flow chart illustrating the different steps in AAMAS.*

Although the initial use of AAMAS is to check specific mixtures for resistance to various forms of distress, the ultimate use, when fully developed and after correlations with field performance of pavements, will be to optimize the structural and mixture design process to produce the desired pavement performance at least cost. The readers should note, however, that some changes to the procedure will likely be made after all testing and field studies have been completed.

18   ASPHALT CONCRETE MIX DESIGN

Factors Considered in the AAMAS Concept

Any mixture design or analysis method should take into account as many of the actual plant production and construction variables as possible and minimize the compromises that must be made between the laboratory procedures and "real life." This mixture analysis system must also take into consideration the behavior of the mixture under traffic and environmental conditioning. However, from the authors' experiences, considerable differences do exist between specimens prepared in the laboratory and samples recovered from the field. Some of these differences are mentioned later to understand how and why a mixture design value or property measured in the laboratory might differ from that obtained from the actual construction process.

Differences Between Laboratory and Plant Production

The asphalt concrete mixture design procedure should duplicate, as closely as possible, the properties and characteristics of the "same" mix manufactured in an asphalt plant. Considerable differences exist between the asphalt concrete manufacturing processes used in the laboratory and those actually used in a batch or drum mix plant. For example, Von Quintus et al. [1] found that air voids were generally higher for plant-mixed material when compared to laboratory-mixed material using the same compactive effort, as shown in Fig. 3 [2-4]. The authors also suggest that significant differences can occur between mixtures produced in different type plants (drum versus batch plants). Some of the critical factors addressed in AAMAS and related to mixture production include a plant hardening simulation, mixing temperature, minus 200 material, maximum aggregate size, and sample size.

Differences Between Laboratory and Field Compaction

The idea of any laboratory compaction process is to simulate, as closely as possible, the actual compaction effort produced in the field by the rollers and traffic. This comparison includes such factors as particle orientation, total air-void content, and void structure (such as number of interconnected voids). Voids are especially important, because of their effect on the engineering properties, which has been substantiated by numerous research and testing studies; for example, Fig. 4 [5].

Some studies have been conducted that compare mixture design values where different laboratory devices were used to compact samples to approximately equal air voids or densities. One such study was conducted by Jiminez [6] for comparing specimens compacted by both the Arizona vibratory/kneading compactor and a Triaxial Institute compactor. Figure 5 illustrates that although densities were equal, Hveem stabilities were significantly different. Another study is currently being conducted by Dallas Little at Texas A&M University to improve mixture design for the Texas State Department of Highways and Public Transportation (SDHPT). In this study, indirect tensile strengths were measured on samples compacted to similar air voids with the Marshall Hammer and Gyratory Shear Compactor (Fig. 6). Consistent differences in indirect tensile strengths were found between specimens compacted with the different devices.

Limited comparisons have also been made by Nunn [7] between samples compacted with a laboratory rolling wheel and modified proctor. Results indicate that use of the modified proctor resulted in large material property variations in the axial and radial directions, and samples were much more resistant to rutting than samples prepared with the rolling

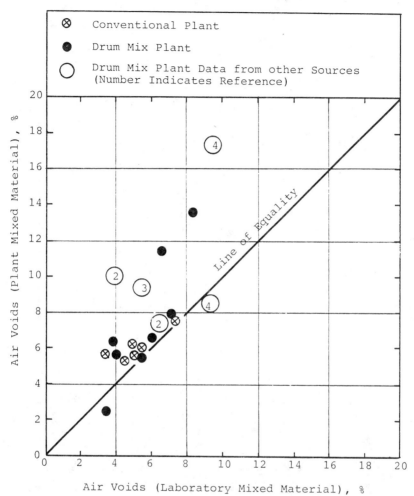

FIG. 3—*Comparison of air voids between laboratory and plant mixed materials for different mixing processes (after Ref 1).*

wheel. Figure 7 illustrates some of the comparisons. The rolling wheel was found to more closely simulate field compaction.

Unfortunately, there have been very limited studies comparing the engineering properties of samples prepared in the laboratory (using different types of compaction equipment) to samples compacted in the field with standard compaction equipment. Based on the few studies performed, the mixture properties will vary with compaction equipment (assuming identical sample size and air voids). This difference may be a result of different particle orientation, or that some of the impact-type compactors fracture the aggregate sooner than contact or roller-type compactors. As a result, field studies are being conducted to determine which type of laboratory-compaction device(s) more closely simulate the engineering

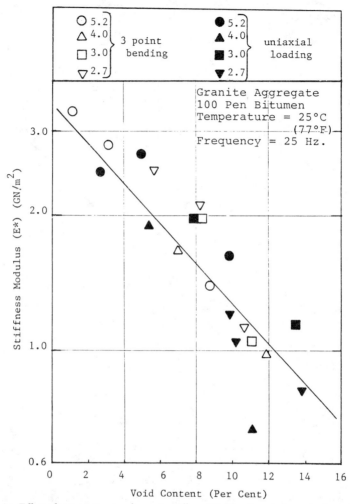

FIG. 4—*Effect of compaction on dynamic stiffness* [5]. *Numbers in legend refer to bitumen content in percent.*

properties of field-compacted specimens. Some of the critical elements related to placement and compaction that must be considered in AAMAS include: particle orientation, maximum aggregate size, sample size, and initial air voids.

Differences Between Initial and Long-Term Conditions

The density obtained in an asphalt concrete lift after the rolling process is not constant when subjected to traffic for a period of time. Many mixtures are designed at an air-void content of 3 to 5% (for example, the Asphalt Institute's MS-2 Manual). However, based on the authors' experience, some asphalt concrete mixtures are placed with air-void contents over 8%, and sometimes even over 10%. These mixtures, particularly those with high air-

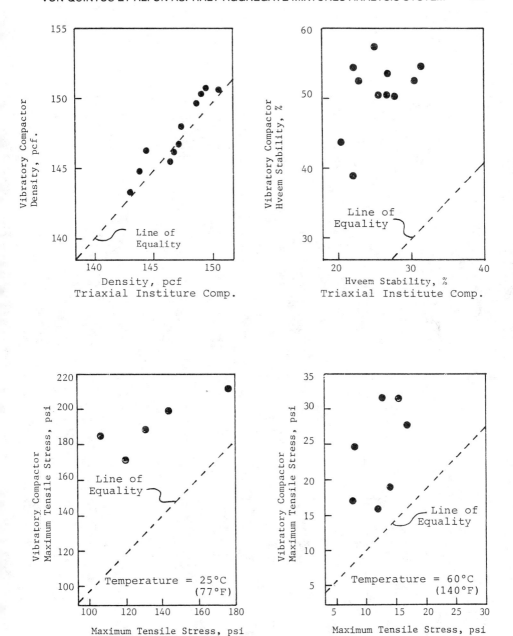

FIG. 5—*Comparison of compaction effects of specimen properties (after Ref 6).*

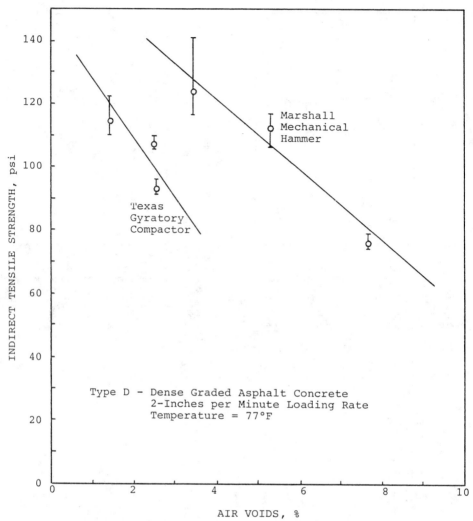

FIG. 6—*Effect of air voids on the indirect tensile strength of laboratory compacted specimens using the Marshall Hammer and Gyratory Shear (unpublished data).*

void levels, will densify further under traffic and become stiffer with time (due to asphalt hardening).

This air-void reduction process, often called rutting, is dependent on a number of mixture design, construction, and environmental variables. In any case, the density value obtained after compaction in the laboratory remains constant. Under traffic, the air-void content of the asphalt concrete mixture decreases with time, which changes the engineering properties of the mixture. Additionally, the initial and change in air voids with time also affect long-term age hardening of the asphalt. Thus, there is a difference in the short-term and long-term air void content of an in-service asphalt concrete mixture. Some of these differences can be dramatic while others are much less, but both the change in air voids and hardening of the mixture must be considered in AAMAS.

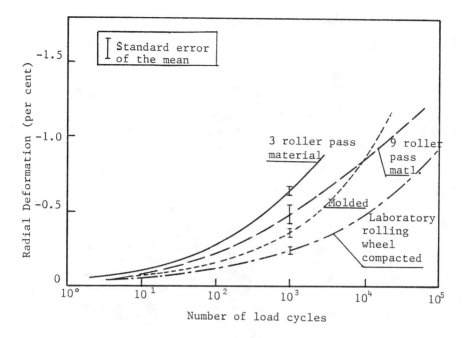
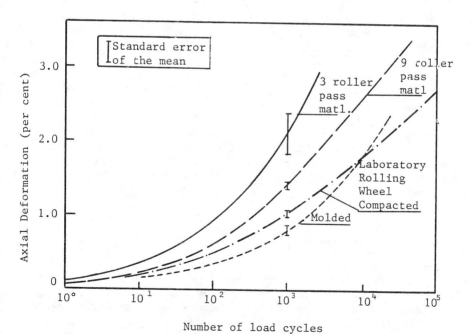

FIG. 7—*Variation of radial and axial deformation with load cycles for samples compacted in the laboratory and field cores (after Ref 7).*

Considerations of Traffic and Environmental Distresses

Potential pavement distress begins as soon as construction ends. In the authors' opinion, traffic is usually a much more severe factor than are environmental conditions. In the case of some distress mechanisms, for example, moisture damage, a combination of the two factors, acts to cause the distress. Neither the Hveem or Marshall mixture design procedure relate material parameters to pavement distress. The AAMAS, however, must address both anticipated traffic levels and environmental conditions (Fig. 2).

As "structural performance" is a rather general description, it is necessary to relate specific pavement performance measures to "functional failure" of flexible pavements. Functional failure implies an unsatisfactory level of serviceability, either in terms of unsatisfactory ride quality, unsafe conditions, or levels of distress that warrant repair or rehabilitation. Most of the original structural design work that has been sponsored as part of Federal Highway Administration (FHWA) and NCHRP projects (for example, Refs 8 and 9) have recognized three basic types of distress. These are: (1) thermal or low-temperature cracking, (2) fatigue cracking, and (3) rutting or permanent deformations. However, other distress types can be equally important, but historically have received much less study. These include stripping or moisture damage, reduced skid resistance, raveling, and bleeding. Asphalt hardening or aging is also very important to long-term pavement performance. Aging is not considered a distress, but a factor that has an extremely important impact on the distresses just listed. Thus, this phenomenon must also be evaluated.

Mixture Evaluation Phases

The laboratory evaluation portion of AAMAS is divided into three basic phases. The first phase is simply the initial mixture design that is conducted with current acceptable mixture design procedures, such as Marshall and Hveem. Once an initial mixture design has been developed, these materials are mixed, compacted, and conditioned in the second phase. This second phase provides an age hardening simulation (both initial and long-term environmental), moisture conditioning and evaluation, and specimen preparation for testing. This phase will be identified as the Mixture Compaction/Conditioning Phase. Once the materials have been mixed, compacted, and conditioned, the specimens are tested and evaluated in the third phase that measures the engineering properties of the mixtures. The engineering properties are used to predict performance (as related to the assumptions used in pavement thickness design), or can be simply compared to acceptance/rejection criteria. This third phase will be identified as the Integrated Mixture Evaluation Phase. Figure 8 shows an overall flow chart for the development of AAMAS.

Mixture Compaction/Conditioning Phase

As previously discussed, considerable differences can exist between asphalt concrete mixtures prepared in the laboratory, as opposed to those that are compacted in the field. The major differences include compaction techniques and procedures, sample/aggregate size, age hardening of the asphalt cement, and moisture conditioning of the asphalt concrete mixture. Ideally, AAMAS should be capable of evaluating the plant-mixed and roadway-compacted properties of the mixture during the mixture design stage.

In order to duplicate, as closely as possible, the properties and characteristics of the in-place mixture, three primary steps are required in the laboratory. These consist of (1) age hardening using accelerated weathering tests to duplicate the hardening of the asphalt

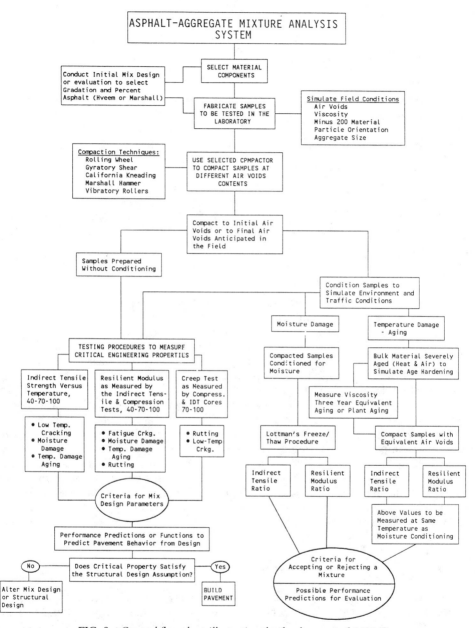

FIG. 8—*General flow chart illustrating the development of AAMAS.*

through an asphalt plant and in the environment with time and traffic, (2) moisture conditioning the asphalt concrete specimens to evaluate the moisture damage of a particular mix, and (3) compacting the materials in the laboratory to obtain the same characteristics of field cores (both initially after compaction and after many millions of load repetitions).

Age Hardening

Two important factors in evaluating how engineering properties in an asphalt concrete may vary with time are initial air voids of the mixture and hardening of the asphalt cement with time. It has been shown by many researchers (for example, Ref *10*), that as penetrations decrease and viscosities increase, the stiffness of the mixture will also increase. This increase in stiffness with time affects not only the fatigue characteristics of the mixture but also the permanent deformation and load temperature cracking potential. Therefore, evaluating and predicting the performance of asphalt concrete with time requires that the asphalt cement during the laboratory evaluation testing phase be placed in an accelerated weathering test.

For the development of AAMAS, six different age-hardening techniques are being used for evaluating the long-term engineering properties of compacted specimen, in cooperation with the Florida Department of Transportation. These are outdoor exposure, 60°C (140°F) convection oven, 60°C (140°F) forced draft oven, 60°C (140°F) with ultra-violet exposure, Oregon's oxygen chamber [*11*], and the thin-film oven test.

Compaction of Laboratory Specimen

There are numerous procedures that can be used to compact laboratory specimens for testing and evaluation. The most critical item required for selecting a laboratory compaction procedure for AAMAS is ensuring that the engineering properties of laboratory-prepared samples are equivalent to the same engineering properties of the in-place material. For the development of AAMAS, field test sections are being constructed initially in Colorado, Michigan, Texas, Virginia, and Wyoming to recover cores of the in-place mix compacted with different compaction trains.

Three different types of breakdown rollers are being used; static steel wheel, vibratory and pneumatic rubber-tired rollers. Plant mixed material from these same projects is being compacted in the laboratory using five devices. These are the Marshall Hammer, the kneading compactor, the Gyratory Shear Compactor, the Arizona Vibratory/Kneading Compactor, and the Mobile Steel Wheel Simulator. Indirect tensile strengths, resilient moduli, creep, and tensile strains at failure will be measured on all laboratory specimens (compacted to air voids measured on field cores) and compared to the same test results on field cores. The compaction procedure(s) that more closely simulates the field cores will be selected for further studies.

Moisture Conditioning

This damage mode has been addressed by Lottman [*12*] and more recently by Tunnicliff and Roof [*13*]. Although the tie-in between specimen condition and performance is not finalized, several of the concepts that these researchers have addressed should be useful in AAMAS to evaluate potential moisture damage. One of the important findings is that laboratory samples must be prepared or compacted to similar air voids in the field to predict potential stripping of some mixtures. As no other procedures were identified as being equal to the amount of research and field verification as conducted by Lottman, his procedures were recommended for AAMAS. The moisture damage results are used for acceptance or rejection of the mixture.

Integrated Mixture Evaluation Phase

There has been an enormous amount of testing that has been conducted in the past relating variations in material properties (asphalt, content, voids, VMA, gradation) to some engineering property or strength value typically required in analytical models, especially using the repeated-load indirect tension test. Indirect tensile strengths, resilient moduli, and fatigue life have been measured and reported on many mixtures by Kennedy and others (for example, Refs *14–16*). Figures 9, 10, and 11 show typical results of tests used to evaluate optimum asphalt contents for various mixtures.

To qualify as an acceptable test to measure the material and engineering properties of a

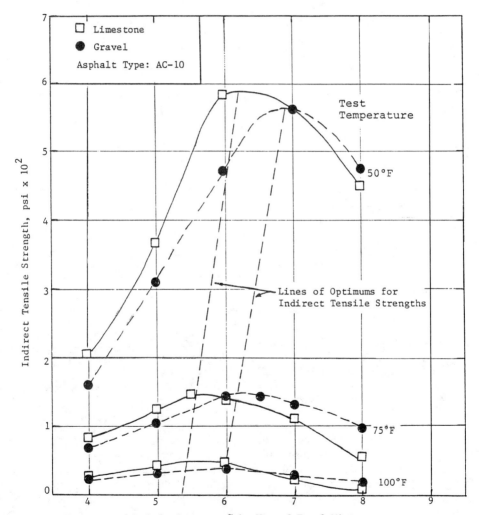

FIG. 9—*Relationships between average indirect tensile strength and asphalt content for limestone and gravel mixtures (after Ref 15).*

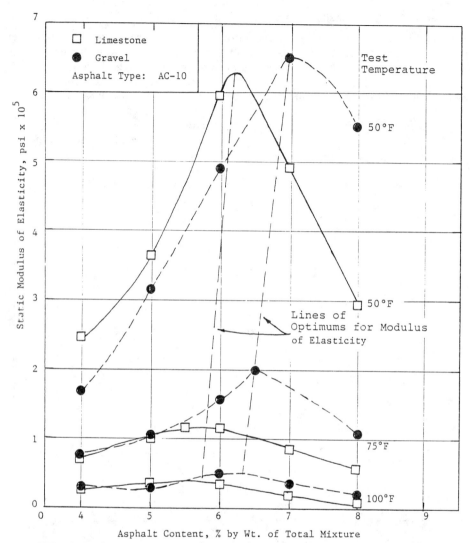

FIG. 10—*Relationship between average static modulus of elasticity and asphalt content for limestone and gravel mixtures (after Ref 15).*

mixture for use in AAMAS, a test must possess the following attributes: (1) reliability or accuracy, (2) repeatability, (3) sensitivity to mixture variables, (4) efficiency of testing, and (5) simplicity of testing. In order to design mixtures based on pavement behavior and performance, it will also be necessary to use test methods that provide or evaluate the necessary engineering properties and characteristics of the asphalt concrete mixture (Fig. 2).

Some of the more common properties that are required to evaluate pavement distress and performance by many mathematical models include resilient modulus, creep compliance, indirect tensile strength, fatigue, and permanent deformation. Based on the authors' experiences, however, there is generally a poor correlation between these engineering properties and typical mixture design values such as Marshall or Hveem stability. Thus, three

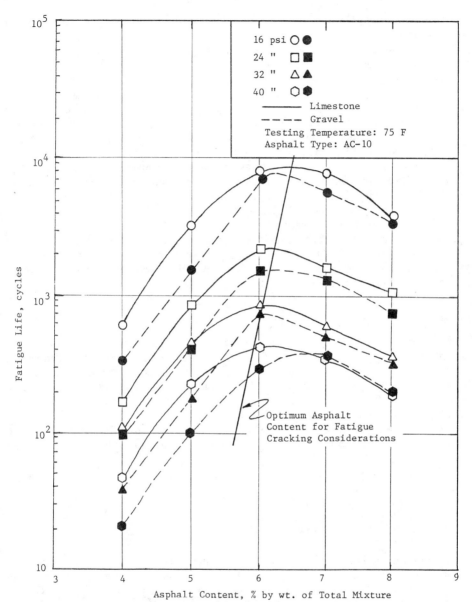

FIG. 11—*Effects of aggregate and asphalt content on fatigue life (after Ref 15).*

basic types of tests are initially being used as tools in the mixture evaluation phase. These tests are: the creep test, the indirect tensile resilient modulus and strength tests, and the repeated load permanent deformation test. These tests are used to evaluate mixture acceptability over a temperature spectrum to which mixtures will be subjected in the field (Fig. 8).

Fatigue tests have also been used to evaluate different mixes, but are very time consuming, require many samples for testing, and the test results are highly variable, at best. However, both Kennedy [17] and Little [18] have developed relationships between creep and fatigue tests for the evaluation of different mixtures. Rauhut et al. [19] also related the fatigue constants, K_1 and K_2, to asphalt concrete resilient modulus and temperature based on numerous data accumulated in the literature. Therefore, the resilient modulus or creep tests or both are initially being used to estimate the fatigue characteristics of asphalt concrete mixtures. Fatigue tests are not being performed in this phase of AAMAS.

During the past ten years, researchers have approximated asphalt concrete behavior under load as if it behaves linearly viscoelastically. Little et al. [20] have demonstrated that the assumption of linear viscoelasticity for asphalt concrete mixtures is acceptable under the types of loading and temperatures normally occurring in highway environments. Employing the principles of linear viscoelasticity allows considerable freedom of manipulation and extension of the stiffness versus time or temperature relationship or both. Specifically, the following principles can be employed:

1. Time-temperature superposition.
2. Approximation of effects of cumulative dynamic loading by static creep tests.
3. Ability to employ time-temperature superposition in order to account for the effects of temperature variation on creep data and hence the effects of temperature variation on repeated load-induced deformation.

Indirect Tension Test

The indirect tension test (IDT) simulates the state of stress in the lower position of the asphalt layer or tension zone. The IDT appears to be a practical and rather simple test method for characterizing asphalt concrete properties or failures caused by tensile stresses. In addition, the test results are not affected by surface conditions; failure occurs in an area of a relatively uniform tensile stress; and the variation of test results is rather low compared to other types of tests.

The indirect tension test can be conducted with a single load to failure or with repeated loads. These two forms of the test provide information related to the following properties:

1. Tensile strength.
2. Poisson's ratio.
3. Resilient modulus of elasticity.
4. Fatigue life (fatigue constants, K_1 and K_2).
5. Permanent deformation (Alpha (α) and Nu (ν) constants).

In addition, the method provides information on moisture susceptibility (Lottman [12]). It also appears that meaningful creep behavior information can be obtained based on the work of Khosla and Omer [16]. Figure 12 summarizes a portion of Khosla and Omer's findings comparing indirect tensile to axial compression compliance data. As shown, for short-loading durations, no significant difference was found, whereas, for long-load durations, a significant difference results. Long-load durations are typical for thermal loadings and short durations typical for wheel loadings. This possibility will minimize or eliminate the need for long-term repeated load tests.

It has been also used by Button et al. [21] to predict and evaluate tender mixes. A portion of Button's findings are shown in Fig. 13. Basically, Marshall stability did not correlate to mixture tenderness, whereas, splitting tensile strain and strength were found to be related to tenderness. Thus, it is possible that a single test, which can be conducted under a variety of test conditions (that is, temperature, moisture conditions, loading rates, etc.), could be

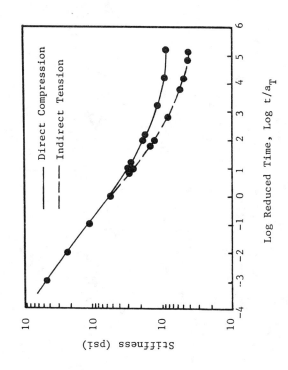

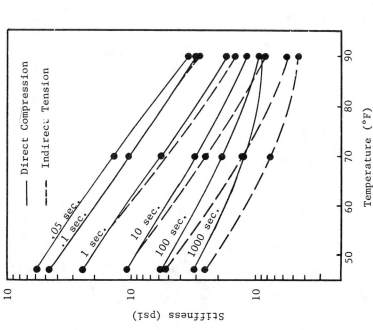

FIG. 12—*Stiffness of asphalt concrete as a function of temperature and load duration (after Ref 16).*

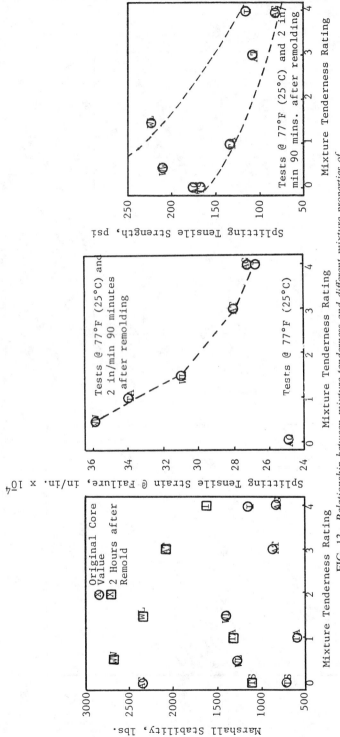

FIG. 13—*Relationship between mixture tenderness and different mixture properties of original cores and remolded mixtures (after Ref 21).*

used to determine the engineering properties needed for elastic or viscoelastic analyses and for evaluating thermal cracking, fatigue cracking, permanent deformation, and moisture susceptibility.

The Creep Test

The creep test can hypothetically be used in the integrated mixture evaluation phase to provide data for two purposes: (1) to evaluate susceptibility to deformation and (2) to determine stiffness at long durations of loading for use in thermal cracking analyses. As presently used, the uniaxial compression creep test typically requires 203-mm-high by 101-mm-diameter (8-in.-high by 4-in.-diameter) specimens. However, research by Shell [22] has shown that cylinders on the order of 101 cm high and 101 cm in diameter (4 in. high and 4 in. in diameter) have provided acceptable data. The assessment on the use of the shorter specimens, however, is a critical research topic.

Evaluation of specimen sizing effects is being conducted by Little in a study entitled "Development of Improved Asphalt Mixture Design Procedures" and sponsored by the Texas Department of Highway and Public Transportation. In fact, research already completed in the Texas study with the closed loop, servo-hydraulically actuated MTS system has demonstrated that a simplified approach to creep testing is not only feasible, but will produce highly reliable results. The Texas study will recommend the use of a simple creep system, which will function similarly to the soil consolidation test apparatus.

Research from the Texas study has also established that compliance versus time of loading and compliance versus temperature relationships are quite sensitive to binder content, aggregate gradation, and air-void content. Because of the sensitivity of compliance to mixture variables coupled with its sensitivity to conditions of loading and temperature, it is possible to establish a band of creep compliance versus either time of loading or temperature that is acceptable for specific climatic and traffic conditions.

At high temperatures or long durations of loading or both, the viscous component of stiffness is predominate. This is the irreversible component that results in permanent deformation. On the other hand, at low temperatures the elastic component of stiffness predominates. The elastic component is immediately recoverable upon the release of the load. The viscoelastic component deforms with time and its recovery is complete but is time dependent. Thus, only the viscous portion is irrecoverable and leads to permanent deformation. This fact may be used with respect to the creep test to predict the magnitude of permanent deformation expected for various mixtures. A clear definition of the effects of nonlinearity is imperative for reliable prediction of performance. Rutting is a nonlinear function of load and tire pressure.

Khosla and Omer [16] have compared IDT compliance data with axial compression compliance data (Fig. 12). At long load durations, a significant difference results. It is at these long load durations (approximately 7200 to 20 000 s) that mixture stiffness should be determined, in order to evaluate induced tensile stresses due to thermal cycling. Thus, it is obvious that AAMAS must also consider the tensile mode of creep, in addition to the compression mode.

Resilient Modulus Test

The resilient modulus of asphalt concrete samples can be measured in a variety of ways. The diametral resilient modulus procedure, introduced on a practical basis by Schmidt in the early 1970s, has been used extensively in asphalt concrete research at Texas A&M University and at the University of Texas. Extensive data bases developed by Epps and Little and Monismith et al. [23,24] and by Little et al. [18,20] have demonstrated that a resilient moduli versus temperature relationship can be used to identify acceptable and unacceptable asphalt concrete mixtures based on their performance. In fact, the concept of performance prediction integrated with full depth pavement design is used by Chevron [25]. The Chevron procedure is based on diametral resilient modulus versus temperature data.

The resilient modulus (MR) versus temperature relationship can be used to ascertain the acceptability of asphalt concrete over the nominal temperature ranges that the pavement will face in the field. Specifically, the MR versus temperature relationship can be used to evaluate the susceptibility of the mixture to cracking and the ability of the mixture to fatigue cracking by the use of the relationship between K_1 and K_2, temperature, and the resilient modulus, as previously discussed.

Repeated load triaxial compression testing with and without confinement were conducted by Von Quintus et al. [26] to measure the resilient moduli of asphalt concrete surface and base layers. The repeated load, unconfined compression, resilient modulus tests were conducted in accordance with ASTM Test Method for Dynamic Modulus of Asphalt Mixtures (D 3497-79). For the confined tests, the sample was encased by a triaxial membrane and air pressure applied within an enlarged triaxial cell, as described by Barksdale [27]. Significant differences in stiffness values were obtained at the higher temperatures (Fig. 14) for the unconfined conditions. Thus, resilient moduli measured at the higher tem-

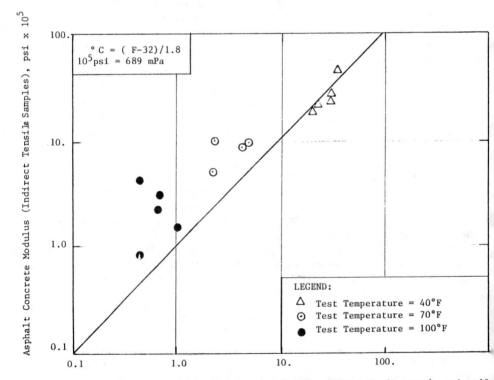

FIG. 14—*Comparison of test results between the unconfined compression and indirect tensile tests.*

peratures using diametral indirect and compression testing techniques are both being used to initially evaluate asphalt concrete mixtures.

Summary

A concept for the asphalt-aggregate mixture analysis system has been presented and is currently under development through NCHRP Project 9-6(1). This AAMAS concept identifies some of the critical forms of pavement distress that asphalt concrete mixtures must resist. In addition, the concept emphasizes the importance of sample preparation (both sample size, void content, and particle orientation), sample conditioning (which includes age hardening, moisture conditioning, traffic densification), and testing methods and configurations to determine the engineering properties for evaluating the behavior and performance of the mixtures.

It is expected that conventional procedures and tests can be used in the development of AAMAS. However, modifications or alterations or both to existing procedures will be required. Laboratory compaction of the asphalt concrete mixtures and the conditioning of those materials prior to testing is a critical and very important item in the development of AAMAS.

References

[1] Von Quintus, H. L., Kennedy, T. W., and Epps, J., "Operational and Performance Characteristics of Drum Mix Plants," Report No. FHWA-TS-83-202, Federal Highway Administration, Nov. 1982.
[2] Apostolos, J. A. and Mann, G. W., "Evaluation of Asphalt Concrete Produced by the Dryer-Drum Mixing Process," Report No. CA-DOT-TL 3125-1-7416, California Department of Transportation, Nov. 1974.
[3] Croteau, J. and Santoro, R., "Evaluation of Bituminous Mixtures Produced by the Dryer-Drum Process," Report No. FHWA-NJ-RD-79-006, Federal Highway Administration, March 1979.
[4] Murphy, K. H., "Production of Asphalt Paving Mixtures with the Dryer-Drum Mixer," Research Report 210, Florida Department of Transportation, May 1979.
[5] Powell, W. D., Lister, N. W., and Leech, D., "Improved Compaction of Dense Graded Bituminous Macadams," *Proceedings*, Association of Asphalt Paving Technologists, Vol. 50, 1981.
[6] Jimenez, R. A., "Structural Design of Asphalt Pavements," Report No. ADOT-RS-13(142), Arizona Department of Transportation, Nov. 1975.
[7] Nunn, M. E., "Deformation Testing of Dense Coated Macadam—Effect of Method of Compaction," Report No. TRLL 870, Transportation Roads and Research Laboratory, 1978.
[8] Finn, F. N., Saraf, C., Kulkarni, R., Nair, K., Smith, W., and Abdullah, A., "Development of Pavement Structural Subsystems," Final Report, NCHRP Project 1-10B, National Cooperative Highway Research Program, Feb. 1977.
[9] "Asphalt Concrete Overlays of Flexible Pavements—Volume I, Development of New Design Criteria," Austin Research Engineers, Inc., Report No. FHWA-RD-75-75, Federal Highway Administration, June 1975.
[10] Witczak, M. W., "Development of Regression Model for Asphalt Concrete Modulus for Use in MS-1 Study," The Asphalt Institute, College Park, MD, 1987.
[11] Ok-Kee Kim, Bell, C. A., Wilson, J. E., and Boyle, G., "Development of Laboratory Oxidative Aging Procedures for Asphalt Cements and Asphalt Mixtures," TRR No. 1115, Transportation Research Board, Washington, DC, 1987.
[12] Lottman, R. P., "Predicting Moisture-Induced Damage to Asphaltic Concrete, Field Evaluation," NCHRP Report 246, National Cooperative Highway Research Program, Transportation Research Board, May 1982.
[13] Tunnicliff, D. G. and Root, R. E., "Use of Antistripping Additives in Asphalt Concrete Mixtures, Laboratory Phase," NCHRP Report 274, National Cooperative Highway Research Program, Transportation Research Board, Dec. 1984.

[14] Kennedy, T. W. and Navarro, D., "Fatigue and Repeated-Load Elastic Characteristics of Inservice Asphalt-Treated Materials," Research Report 183-2, Center for Highway Research, The University of Texas at Austin, Jan. 1975.
[15] Kennedy, T. W., Gonzalez, G., and Anagnos, J. N., "Evaluation of the Resilient Elastic Characteristics of Asphalt Mixtures Using the Indirect Tensile Test," Research Report 183-6, Center for Highway Research, The University of Texas at Austin, Nov. 1975.
[16] Khosla, N. P. and Omer, M. S., "Characterization of Asphaltic Mixtures for Prediction of Pavement Performance," TRR No. 1034, Transportation Research Board, 1985.
[17] Kennedy, T. W., Vallejo, J., and Haas, R., "Permanent Deformation Characteristics of Asphalt Mixtures by Repeated-Load Indirect Tensile Test," Research Report 183-7, Center for Highway Research, The University of Texas at Austin, June 1976.
[18] Little, D. N. and Richey, B. L., "A Mixture Design Procedure Based on the Failure Envelope Concept," *Proceedings,* Association of Asphalt Paving Technologists, Vol. 52, 1983.
[19] Rauhut, J. B., Lytton, R. L., and Darter, M. I., "Pavement Damage Functions for Cost Allocation, Volume 2, Descriptions of Detailed Studies," Report No. FHWA/RD-84/019, prepared by Brent Rauhut Engineering Inc. for the Federal Highway Administration, June 1984.
[20] Little, D. N., et al., "Sulphlex Engineering Properties," FHWA Report FHWA/RD-85/032, Federal Highway Administration, 1985.
[21] Button, J. W., Epps, J. A., Little, D. N., and Gallaway, B. M., "Influence of Asphalt Temperature Susceptibility on Pavement Construction and Performance," NCHRP Report 268, National Cooperative Highway Research Program, Transportation Research Board, 1984.
[22] Claessen, A. I. M., Edwards, J. M., Sommer, P., and Uge, P., "Asphalt Pavement Design—The Shell Method," *Proceedings,* Fourth International Conference on Structural Design of Asphalt Pavements, Vol. 1, Aug. 1977.
[23] Epps, J. A. and Little, D. N., "Mixture Properties of Recycled Control Plant Materials," *Recycling of Bituminous Pavements, ASTM STP 662,* L. Wood, Ed., American Society for Testing and Materials, Philadelphia, 1978.
[24] Monismith, C. L., Epps, J. A., and Finn, F. N., "Improved Asphalt Mix Design," *Proceedings,* Association of Asphalt Paving Technologists, Vol. 54, 1985.
[25] Santicci, E. L., "Thickness Design Procedures for Asphalt and Emulsified Asphalt Mixes," *Proceedings,* Fourth International Conference on the Structural Design of Asphalt Pavements, University of Michigan, Ann Arbor, Vol. 1, 1977.
[26] Von Quintus, H. L., Rauhut, J. B., and Kennedy, T. W., "Comparisons of Asphalt Concrete Stiffness as Measured by Various Testing Techniques," *Proceedings,* Association of Asphalt Paving Technologists, Vol. 47, 1978.
[27] Barksdale, R. D., "Practical Application of Fatigue and Rutting Tests on Bituminous Base Mixes," *Proceedings,* Association of Asphalt Paving Technologists, Vol. 47, 1978.

Discussion

J. L. McRae[1] *(written discussion)*—The compaction phenomenon dictates the requirement that the pavement mixture design test be based upon the ultimate pavement density. This is clearly illustrated in Fig. 15 where the compaction optimum for 1500 coverages under 200 psi tires is at 1% less asphalt and 3 lb/ft^3 higher density than the optimum for construction compaction. I do not believe the authors' approach of basing the laboratory design test on the density achieved under conventional construction equipment will meet this essential requirement. The compaction used in the laboratory design test should relate rationally to the anticipated unit pressure and intensity of traffic. This requires compacting under the design stress essentially to the asymptote of density versus gyrations in the Gyratory Testing Machine (GTM)1. Quoting from "Effect of Aggregate on Performance of Bituminous Concrete" by Brown, McRae, and Crawley in *ASTM STP 1016, Implication of Aggregates in the Design, Construction, and Performance of Flexible Pavements:*

---

[1] Engineering Developments Co., Inc., Vicksburg, MS 39180.

DISCUSSION ON ASPHALT-AGGREGATE MIXTURES ANALYSIS SYSTEM

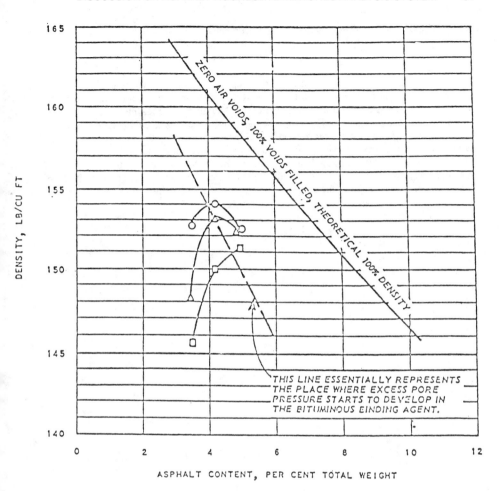

FIG. 15—*Traffic compaction (U.S. Army Engineer Waterways Experiment Station Traffic Test Section).*

In the light of current knowledge not only must the ultimate density be produced in defining the optimum condition but it should be done by a kneading process employing the anticipated pavement design stress.[2] This is because the internal particle structural orientation and arrangement is a function of the compaction process and the structural arrangement of the compacted aggregate has a profound effect on the stress-strain properties of the compacted mix.[2,3] This requires an acceptable mechanical

---

[2] McRae, J. J., "Theoretical Aspects of Asphalt Concrete Mix Design," *Proceedings*, Third Paving Conference, Civil Engineering Department, The University of New Mexico, Albuquerque, 1965.

[3] Monismith, C. L. and Vallerga, B. A., "Relationship Between Density and Stability of Asphalt Paving Mixtures," *Proceedings*, Association of Asphalt Paving Technologists, Vol. 47, 1978.

analog of the pavement in the laboratory test. Ironically, Marshall's basic concept of requiring the ultimate pavement density for the laboratory test still holds while his method of test has not been able to keep abreast of the modern requirements for density and simulation of the stress-strain properties of the prototype.

McRae[2] has advanced the hypothesis that the optimum condition is represented by a statistical equilibrium between the applied load (anticipated design stress) and the body forces of the particles, and this concurrently at a minimum internal energy situation, close to, but slightly greater than, the bitumen content at which the maximum unit weight of the aggregate only occurs. It is believed that the GTM employing kneading compaction under the anticipated design stress and employing the gyrograph to sense plasticity identifies this condition for all practical purposes.

When employing the anticipated design stress in the GTM, widening of the Gyrograph identifies this condition with due regard for the voids in the mix, as well as the plasticity and shear properties, thus avoiding the necessity of the empirical correlation with the voids.

C. L. Monismith,[1] *F. N. Finn,*[2] *and B. A. Vallerga*[3]

A Comprehensive Asphalt Concrete Mixture Design System

REFERENCE: Monismith, C. L., Finn, F. N., and Vallerga, B. A., "**A Comprehensive Asphalt Concrete Mixture Design System,**" *Asphalt Concrete Mix Design: Development of More Rational Approaches, ASTM STP 1041*, W. Gartner, Jr., Ed., American Society for Testing and Materials, Philadelphia, 1989, pp. 39–71.

ABSTRACT: The design of an asphalt concrete mix consists, essentially, of the following steps:

1. select type and gradation of aggregate,
2. select type and grade of asphalt, with and without modifier, and
3. select proportionate amount of asphalt in asphalt/aggregate blend.

These steps have been incorporated into a general framework for design, which serves as the basis for the mix design procedure presented in this paper.

Essentially, the system consists of a series of subsystems in which the mix components and their relative proportions are selected in a step-by-step procedure to produce a mix that can then be tested and evaluated to ensure that it will perform adequately in the specific pavement section for which it has been formulated. The latter evaluation phase includes the influence of environmental factors, effects of traffic, and the consequence of the anticipated structural cross-section design at the designated site in the following distress modes: fatigue, rutting, thermal cracking, and raveling.

The paper includes a brief discussion of the important factors associated with the various steps of the design process and recommended test procedures to be followed. Information and data are presented to illustrate a procedure that might be followed to ensure that the laboratory method of specimen preparation and fabrication provides specimens with stress versus strain characteristics similar to those obtained by field compaction. The use of the creep test to assist in the design process is illustrated. Some techniques to reduce the amount of testing associated with the methodology described herein are also described.

KEY WORDS: asphalt concrete mix design, fatigue, rutting, thermal cracking, raveling, fracture characteristics, water sensitivity, kneading compaction, creep testing, mix design, asphalt concrete, asphalt specifications

In general terms, the design of an asphalt-aggregate mixture system consists of the following basic steps:

1. select the type and gradation of the mineral aggregate ingredient.
2. select the type and grade of asphalt binder, with or without modification.
3. select the amount of asphalt binder to satisfy the project-specific requirements for mix properties.

---

[1] The Robert Horonjeff Professor of Civil Engineering and research engineer, The Institute of Transportation Studies, University of California, Berkeley, Berkeley, CA 94720.
[2] Senior vice president, ARE, Inc., Scotts Valley, CA 95066.
[3] Consulting civil engineer, B. A. Vallerga, Inc., Oakland, CA 94612.

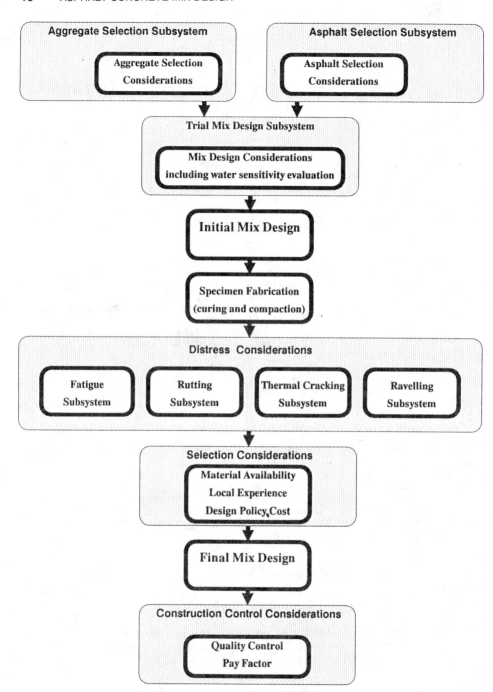

FIG. 1—*A comprehensive design system for asphalt concrete with or without modified asphalt.*

These steps have been incorporated into a general framework for design, as illustrated in Fig. 1, that serves as the basis for the comprehensive asphalt concrete mixture design system presented in this paper.

Proper selection of the mix components and their relative proportions, that is, asphalt or "binder" content, requires a knowledge of the significant properties and performance characteristics of asphalt paving mixtures and how they are influenced by the mix components. Table 1 contains a listing of the mix properties that must be considered for specific design situations together with a summary of the factors that influence these properties.

Mix design is the selection of the components to achieve a desirable balance in these properties for the specific pavement application. As will be seen subsequently, selection of

TABLE 1—*Mixture properties.*

Property	Definition	Examples of Mix Variables that Have an Influence
Stiffness	$S_{mix}(t, T) = \sigma/\epsilon$ Relationship between stress and strain at a specific temperature and time of loading	aggregate gradation asphalt stiffness degree of compaction water sensitivity asphalt content
Stability	resistance to permanent deformation (usually at high temperatures and long times of loading—conditions of low S_{mix})	aggregate surface texture aggregate gradation asphalt stiffness asphalt content degree of compaction water sensitivity
Durability	resistance to weathering effects (both air and water) and to the abrasive action of traffic	asphalt content aggregate gradation degree of compaction water sensitivity
Fatigue resistance	ability of mix to bend repeatedly without fracture	aggregate gradation asphalt content degree of compaction asphalt stiffness water sensitivity (Note: selection of mix components dependent on structural pavement section design.)
Fracture characteristics	strength of mix under single tensile stress application	aggregate gradation aggregate type asphalt content degree of compaction asphalt stiffness water sensitivity
Skid resistance (surface friction characteristics)	ability of mix to provide adequate coefficient of friction between tire and pavement under "wet" conditions	aggregate texture and resistance to polishing aggregate gradation asphalt content
Permeability	ability of air, water, and water vapor to move into and through mix	aggregate gradation asphalt content degree of compaction

the components and their relative proportions is also influenced by the pavement structural section into which the mix will be incorporated. This should make the designer cognizant of the fact that mix design and pavement design are interactive and, therefore, must be considered together.

General Framework for a Comprehensive System

The general framework for the proposed comprehensive design system is shown in Fig. 1. Essentially, the system consists of a series of subsystems in which the mix components, asphalt (or binder) and aggregate, and their relative proportions are selected in a step-by-step procedure to produce a mix that can then be tested and evaluated to ensure that it will attain a desired level of performance in the specific pavement section in which it is to function. The influence of environmental factors, the effects of traffic loading, and the consequence of the pavement structural section design at the selected site is also included in this evaluation.

It is important to note in Fig. 1 that an evaluation for water sensitivity of the mix is scheduled in the trial design subsystem. Satisfactory resolution of this problem prior to examination of the response of the trial mix to the four modes of distress shown in Fig. 1, that is, fatigue, rutting, thermal cracking, and raveling, will allow full concentration on these evaluations.

Depending on the conditions of exposure to climatic and loading factors to which the specific pavement is to be subjected, any or all of the distress modes may be evaluated. For example, in a hot, dry climate, it may not be necessary to examine the potential for thermal cracking whereas, because of the potential for fatigue, rutting, and surface raveling associated with reduced stiffness and asphalt embrittlement, it would be essential to evaluate these latter three modes.

It should be noted that provision has been included in this asphalt aggregate design system to "temper" the design by such factors as material availability, local experience, cost, and risk options associated with the facility and resulting from specific distress modes.

Finally, the system must include provision for considerations of construction control. Levels of compliance must be established to ensure that the mix will achieve the desired performance objective to the requisite level of reliability. Assessment of the influence of noncompliance can also be determined, and appropriate pay factors associated with different levels of nonconformance established.

Aggregate Selection Subsystem

The aggregate selection subsystem is outlined in Fig. 2. The following list includes the aggregate characteristics that should be considered.

1. *Surface Texture*—provides the load-carrying capability of an aggregate through its frictional resistance [1] that, for many aggregates, can be improved by crushing. Surface texture also contributes to the skid resistance of the pavement surface (microtexture [2]).
2. *Durability*—covers the resistance of an aggregate to degradation (production of non-plastic fines), disintegration (production of plastic fines), crushing, polishing, and freeze-thaw.
3. *Wettability*—refers to the propensity of the aggregate to be preferentially wetted by asphalt in the presence of water.

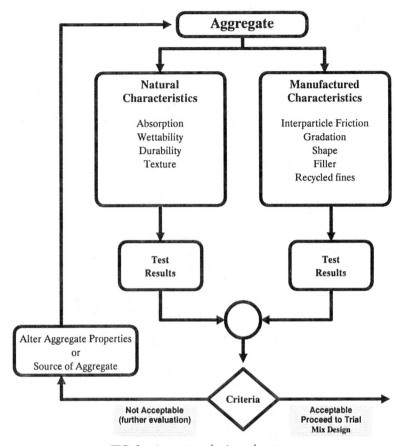

FIG. 2—*Aggregate selection subsystem.*

4. *Size and size distribution*—influences a number of the mix characteristics such as pavement surface texture, permeability, and workability. The maximum aggregate size depends on the thickness of the lift in which the mixture is placed.
5. *Absorption*—is the tendency of asphalt to flow into the pores of the aggregate that affects the amount of asphalt required.
6. *Shape*—may have an influence on the properties of the mixture, for example, stability, creep, and fatigue.

Finally, the type and amount of mineral filler (material passing the No. 200 sieve) can have a significant influence on the properties of the resultant mix (for example, Refs *3-7*).

Depending upon the use of the aggregates, some or all of these characteristics will be considered in the design process. Similarly, requisite criteria for tests adopted to define these characteristics will be dependent on the aggregate's use. Table 2 has been prepared to illustrate, in a general way, how the aggregate use, requisite properties, test methods to define these properties, and criteria can be brought together as part of the aggregate selection subsystem.

Conformance of the aggregate to the selection criteria indicates its acceptability for use

TABLE 2—*Aggregate selection guidelines.*[a]

Properties	Test	Criteria (reference)[b]
1. Friction	(*a*) Percent crushed (CALTRANS 205) (*b*) Texture (ASTM D 3398-87)	AASHTO Guide Specifications 1978
2. Durability	(*a*) Crushing strength	Load to produce 10% or
Crushing resistance	(*b*) ASTM crushing	more fines
Degradation	(*a*) Los Angeles Rattler (ASTM C 131-81 and ASTM C 535-81)	ASTM D 692-88
Disintegration	(*a*) Aggregate Durability Index (ASTM D 3744-85)	
Polishing	(*a*) Accelerated polishing of aggregates using the British Wheel (ASTM D 3319-83) (*b*) Pendulum test (ASTM E 303-83)	
Freeze/thaw	(*a*) Soundness (ASTM C 88-83)	ASTM D 692-88
3. Wettability (preferential)	(*a*) California film stripping test (CALTRANS 302) (*b*) ASTM D 1664-80	
4. Absorption (specific gravity)	(*a*) Specific gravity tests (ASTM C 127-84 and ASTM C 128-84) (*b*) Centrifuge kerosine and oil equivalent tests	
5. Gradation	(*a*) Sieve and wash analysis (ASTM C 136-84a and ASTM C 117-87)	ASTM D 3515-84
	(*b*) Sand equivalent (ASTM D 2419-74)	CALTRANS: 35 to 50%
6. Shape	(*a*) Index of particle shape and texture (ASTM D 3398-87)	
7. Minerology	(*a*) Petrographic analysis (ASTM C 295-85)	

[a] Specific criteria for aggregate will be based on use, that is, highways, airfields, etc.
[b] Specific criteria are not always available for each test and will depend on user criteria.

in the trial mix design phase. If the criteria are not met, modifications may be required as shown in Fig. 2. The process would then be repeated until a suitable material is selected.

Asphalt Selection Subsystem

A framework for the asphalt selection subsystem is shown in Fig. 3. Like the aggregate selection system, the requisite characteristics for the asphalt must be defined for the specific end use of the mix.

Characteristics include both physical and chemical properties. While chemical requirements now appear primarily in specifications for restoring (recycling) agents, results of the Strategic Highway Research Program (SHRP) may lead to such considerations for asphalt and modified asphalt binders as well [*8*].

Physical Properties

1. Rheological Properties—The response of asphalt mixtures to both load and environmental influences is dependent on the rheological characteristics of the asphalt contained therein. These characteristics may be defined by stress-strain relationships and viscosity

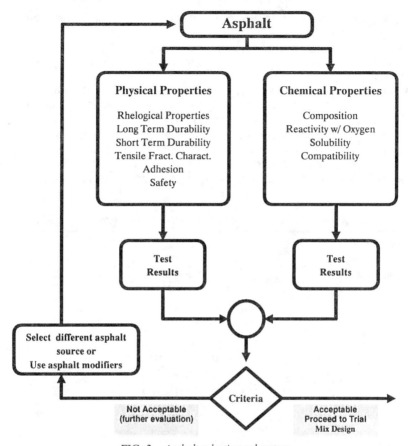

FIG. 3—*Asphalt selection subsystem.*

over a range in temperatures and times of loading. For example, the stiffness modulus of a mixture at a particular time of loading and temperature is a function of the stiffness modulus of the asphalt contained in the mixture [9–11].

2. *Durability*—The resistance of the asphalt to weathering or aging (that is, embrittlement) including both short term (that is, during the hot mix process) and long term (that is, the in-service period). Weathering, which involves such chemical and physical actions as oxidation and volatilization, results in a change in the rheological properties of the asphalt that are generally believed to accelerate the occurrence of most types of distress in asphalt concrete; for example, raveling, low-temperature cracking and fatigue cracking. Weathering may be beneficial, however, when related to rutting and possibly water sensitivity (and possibly to fatigue in thick asphalt concrete sections).

3. *Tensile (Fracture) Characteristics*—The tensile strength of the mix is dependent on the tensile (fracture) strength of the asphalt matrix contained in the mix.

4. *Adhesion*—The bonding characteristics of the asphalt to the aggregate that influences the water sensitivity of the mix.

5. *Safety*—The temperature at which the asphalt will flash should be defined; normally, the temperature at which this occurs for asphalt cements is above the mixing temperature.

Information about the health influences of the asphalt may also be desirable to ensure proper handling of the material.

As with the aggregate, specific properties will be required depending on the end use of the material. Table 3 has been prepared to illustrate the form that such guidelines might take. Included are listings of the properties, potential tests to define the properties, and a few examples of criteria that might be applied to the test results to define a specific end use. Criteria for most of the tests shown in Table 3 are not available. It is hoped that some will be developed as a part of the SHRP endeavor.

Conformance of the test results with the selected criteria would indicate the suitability of the asphalt for use in the trial mix design phase. If the test results do not conform with the criteria, either an asphalt from a different source may be evaluated if that is feasible, or the asphalt characteristics may be improved by the addition of a modifier. A number of

TABLE 3—*Asphalt cement (with and without modifier) selection guidelines.*

Properties	Test	Criteria[a]
1. Rheological properties viscosity	(a) Penetration (ASTM D 5-86) (b) Cone and plate (ASTM D 3205-86) (c) Kinematic (ASTM D 2170-85) (d) Sliding plate (ASTM D 3570-77) (e) Vacuum capillary (ASTM D 2171-85)	
stiffness	(a) Sliding plate rheometer (b) Microelastometer (c) Fraas	
shear susceptibility	(a) Sliding plate (ASTM D 3570-77) (b) Schweyer rheometer	
temperature susceptibility	(a) Viscosity tests at two or more temperatures (b) Stiffness tests at two or more temperatures	
2. Durability during mixing (short term)	(a) Thin film oven (ASTM D 1754-87) (b) RTFO (ASTM D 2872-85)	
in-service (long term)	(a) Tilt oven asphalt durability (TOAD) test (b) Aging index	100 000 P at 140°F
3. Tensile fracture characteristics	(a) Force ductility (tensile elongation) (b) By calculation (Shell procedure)	
4. Adhesion	(a) Film stripping (CALTRANS) (b) Boiling water	<25% <5%
5. Safety	(a) Flash point	

6. Chemical parameters	(a) Percent saturates (ASTM D 4124-86)
	(b) Functional ratios
	  1. Corbett
	  2. Rostler
	  3. Gzemski
	  4. Gotalski
	(c) Purity
		15 to 20%
		Rostler 0.8 to 1.2

[a] Specific criteria have not been established for each test, research under SHRP should establish performance based recommendations.

modifiers are available [12]; the choice of a specific modifier will be dependent upon the characteristics requiring improvement and the cost of modification.

Trial Mix Design Subsystem

The purpose of the trial mix design process (Fig. 1) is to select, for the specific aggregate, aggregate gradation, and type and grade of asphalt, an asphalt content that will serve as a basis for determination of the initial mix design asphalt content. This procedure can be accomplished using some form of computational procedure combined with data from simple and routine type test methods. For example, an asphalt content can be estimated from the following expression

$$P_{W_{asp}} = (SA)(t)(\gamma_{asp})(100) \qquad (1)$$

where

$P_{W_{asp}}$ = asphalt content in percent, by weight of aggregate;
SA = surface area of aggregate, for example, square feet of surface per pound of aggregate (m^2/kg);
t = average film thickness of asphalt reflecting surface texture and absorption characteristics of aggregate, ft (mm); and
γ_{asp} = unit weight of asphalt.

Various procedures have been developed to determine a preliminary asphalt content by this methodology. Reference 13 illustrates one such example that might be followed.

Alternatively, as Nijboer has suggested [14], an estimate can be made of the voids in the mineral aggregate at stable packing of the asphalt-aggregate mixture that, in turn, permits selection of an asphalt content. In this case, the volume of asphalt sufficient to leave some nominal value of air voids such as 3 or 4% (by volume of the mix) can be determined.

Initial Mix Design

The asphalt content selected in the trial mix design subsystem will serve as the basis for the preparation of laboratory specimens for testing in this phase of the process. Mixes containing a range in asphalt contents, bracketing the trial value, should be prepared. This preparation phase is of extreme importance to ensure that the specimens are representative of those produced in the field.

A standardized procedure must be followed for the blending and mixing of the aggregate and asphalt under laboratory conditions and should include criteria for batch size, blending sequence, mixing temperature, and mixing time. Equipment necessary for this standard procedure should also be specified with respect to critical dimensions and specific features needed to ensure consistent and uniform results. A standardized conditioning procedure is required for "curing" the asphalt mixture in the laboratory to permit absorption of the asphalt by the aggregate. Available evidence suggests that such conditioning should be done *prior* to compaction while the mix is in a loose state.

Laboratory compaction is a critical part of this process in that it is imperative that the compaction equipment utilized, for example, gyratory, kneading, and Laboratorie Central Ponts et Chaussees (LCPC) rubber tire [15], produce specimens for laboratory testing that exhibit similar characteristics of those compacted in the field and under specific loading conditions. The compaction method should be able to produce specimens with the following features.

ASPHALT CONCRETE MIX DESIGN

(*a*) Densities (air voids and VMA) corresponding to those desired at various levels of traffic usage, from initial construction (that is, no traffic) to ultimate design traffic (for example, 20 years of heavy-duty traffic).

(*b*) Engineering properties corresponding to those measured by tests on cores taken from the in-situ asphalt paving at various levels of traffic usage.

Later in this paper, data are presented illustrating comparisons between the creep behavior of laboratory-compacted and field-compacted specimens and is representative of the type of data that are, we believe, required to establish the validity of a specific laboratory compaction procedure.

It is also important to emphasize that the compaction method selected has the capability of producing realistic test specimens in the shapes (that is, other than cylindrical) and sizes required for the particular test methods to be specified. It should also include specific details on how to determine, attain, and maintain (that is, during compaction) the compaction temperature.

Some form of the mixture testing must then be accomplished to permit selection of the initial design asphalt content. The "screening" test or tests will be dependent on the use of the mixture. Table 4 provides an indication of tests that might be considered, based on required characteristics and available equipment and procedures. It is clear that all of the tests in Table 4 are not screening tests; however, it will be important to establish a set of tests as indicators in order to have reasonable assurance that further testing is justified. In general, this selection process will require that the designer recognize that a balance among significant properties is required when selecting the asphalt content, for example, Fig. 4.

When the initial mix design has been selected, it should then be tested for water sensitivity. One methodology that might be followed is as follows.

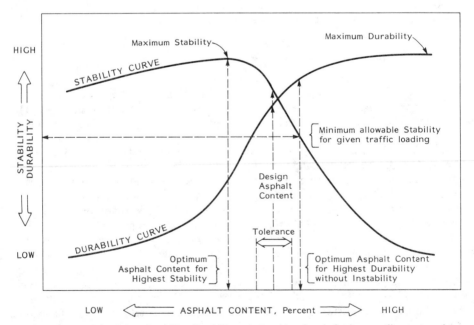

FIG. 4—*Schematic of stability-durability relationship of asphalt concrete illustrating philosophy of selecting the design asphalt content.*

TABLE 4—*Asphalt concrete selection guidelines.*

Properties	Test	Criteria[a]
1. Stiffness (temperature and time of loading) Flexibility	(a) Dynamic (ASTM D 3497-79) (b) Diametral (ASTM D 4123-82) (c) Flexural (d) Creep (Shell, etc.)	
2. Stability (including permanent deformation)	(a) Hveem Stabilometer (ASTM D 1560-81a) (b) Marshall apparatus (ASTM D 1559-82) (c) Triaxial compression test (ϕ, cohesion) (d) Compressive strength of bituminous materials (ASTM D 1074-83)	
3. Durability	(a) Voids analysis (ASTM D 3203-83) (b) Accelerated aging (c) Film thickness	
4. Fatigue characteristics	(a) Flexural (b) Diametral	
5. Fracture characteristics	(a) Split tension (b) Direct tension	
6. Thermal characteristics	(a) Coefficient of expansion (b) Coefficient of thermal conductivity (c) Specific heat capacity	
7. Surface friction characteristics	(a) British pendulum (ASTM E 303-83)	
8. Coefficient of permeability	(a) Water (b) Air	
9. Wear resistance	(a) Surface abrasion of compacted bituminous materials (CALTRANS 360)	
10. Volume change	(a) Absorption (b) Thermal (c) Aging (d) Water (freezing)	
11. Water sensitivity	(a) Bituminous coated aggregate (ASTM D 1664-80) (b) Effect of water on cohesion of compacted bituminous mixtures (ASTM D 1075-81 and ASTM D 1074-83) (c) Lottman procedure (d) Pedestal test (e) MVS (CALTRANS 307) (f) Swell	
12. Asphalt absorption	(a) Percent asphalt absorption by aggregate (ASTM D 4469-85)	
13. Specific gravity and air voids	(a) Theoretical maximum (ASTM D 2041-78) (b) Measured (ASTM D 1188-88 and ASTM D 2726-88) (c) Percent air voids (ASTM D 3203-83)	

[a] Specific criteria are dependent on user organization.

(a) Compact the laboratory specimens to air void contents that span the range of those likely to be obtained in the field. A range of air voids between 4% to as much as 12% is suggested.

(b) The laboratory specimens of similar air-void contents should be divided into two subsets, one to remain in a dry unconditioned state as the control set and the other to be conditioned using the Lottman procedure [16] and subjected to multiple freeze-thaw cycles.

(c) After each freeze-thaw cycle, the retained resilient modulus ratio will be determined by dividing the resilient modulus for the conditioned specimen by the resilient modulus measured initially on the control set.

(d) The resilient modulus test should be conducted at test temperatures and at frequencies stated in ASTM Method for Indirect Tension Test for Resilient Modulus of Bituminous Mixtures (D 4123-82). Both instantaneous and total modulus values should be recorded.

(e) The number of freeze-thaw cycles required is expected to vary according to the environment to which the pavement will be exposed in the field. For the initial testing program, a minimum of five cycles of testing is recommended.

From this evaluation, the mix will either be considered suitable for further testing or determined to be water sensitive. If the mix is water sensitive, it will be necessary to modify the asphalt or the aggregate (or both) such that the mixture will test satisfactory with regard to water sensitivity. With the resulting mixture selected from this process, additional testing can then be accomplished to examine the potential for specific forms of distress depending on the anticipated site conditions.

Distress Considerations

As seen in Fig. 1, the major forms of distress to be examined include: fatigue, rutting, thermal cracking, and raveling. Each of the subsystems will be considered separately. It should be emphasized at the outset that in this process the performance of the mix in the specific pavement structure in which it will be used must be considered in the mix design process. This is illustrated in Fig. 5. For example, if the asphalt proposed for the project is known to have relatively poor durability properties, but still within the specifications, it may be desirable to use a thick layer of asphalt concrete, that is, 152.4 mm (6 in.) or more, in order to benefit the fatigue life of the pavement structure. Thus, pavement design and asphalt mix design must be considered together. This is reflected not only at this stage in the mix design process but in the asphalt, aggregate, and trial mix design subsystem as well. The following sections of this paper describe the steps to be followed when considering each of the distress modes according to the framework of Fig. 5. For the raveling mode of distress, some of the steps shown in this figure are not required.

Fatigue Subsystem

Figure 5 illustrates the essential elements required to consider the fatigue mode of distress in a specific pavement section.[4] The requisite mixture characteristics are: (1) stiffness moduli at short times of loading and in the range of the temperatures expected at the site; and (2) fatigue characteristics.

---

[4] In this case, the distress characteristics would be the fatigue response of the asphalt concrete.

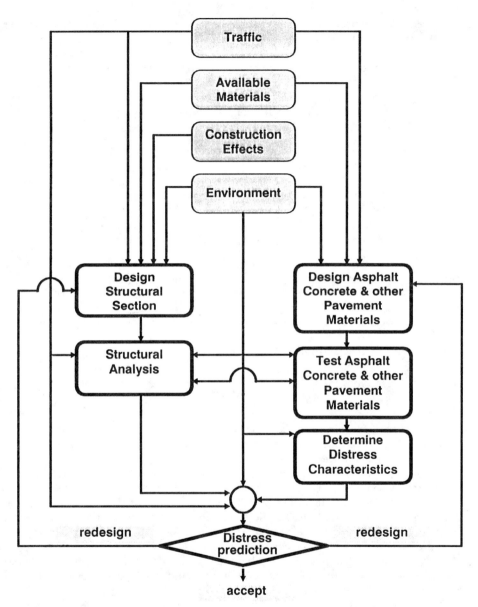

FIG. 5—*Framework of design subsystem to estimate potential for occurrence of particular distress mode.*

Representative pavement temperatures can be estimated to any degree desirable [*17*] to define the pavement stiffness conditions. For a trial pavement thickness, strains in the asphalt bound layer can be determined for the expected range in loads and for representative conditions of stiffness of the asphalt bound layer together with representative stiffnesses of the other components of the pavement as well. Multilayer elastic analysis is recommended, for example, use of the ELSYM [18] or BISAR [19] programs for the computations of stresses, strains, and deflections in multilayer elastic systems.

For specific conditions of mix stiffness, strain values for various axle loads can then be ascertained permitting, in turn, the potential for fatigue cracking to be estimated. This requires an estimate of the repetitions, n_i, for each of the loads anticipated to be applied to the pavement. For each of the strain levels, the expected number of repetitions to cracking, N_i, can be obtained from an equation of the form

$$N_i = k \left(\frac{1}{\varepsilon_t}\right)^n \left(\frac{1}{S_{\text{mix}}}\right)^a \qquad (2)$$

where

ε_t = strain repeatedly applied, in./in. (mm/mm);
S_{mix} = mix stiffness, psi (Pa); and
k, n, a = coefficients experimentally determined.

In many design situations, definition of the fatigue characteristics of asphalt concrete, a costly process, may not be warranted. Accordingly, it will be necessary to have some other measure of mixture response that can be related to the mix fatigue characteristics. A procedure like that used by the LCPC of France [15], in which the tensile characteristics of the mix have been correlated with fatigue response, may be a worthwhile approach.

With some form of Eq 2, damage can be estimated using the linear sum of the cycle ratios cumulative damage hypothesis

$$\sum_{i=1}^{n} \left(\frac{n_i}{N_i}\right) \leq 1 \qquad (3)$$

If the sum of the cycle ratios is close to unity, the design may be considered adequate. If, on the other hand, the sum exceeds or is much less than unity, a new trial pavement section should be selected and the analysis repeated.

For highway pavements, it may be desirable to reduce the number of computations. This can be done by converting repetitions of the various axle loads using equivalency factors such as those developed by the American Association of the State Highway and Transportation Officials (AASHTO) [20]. Tensile strains for the standard axle load can then be determined for the expected range of stiffness conditions of the pavement components, and the sum of the cycle ratios can be determined as just described.

It should also be noted that a design alternative is to consider a change in mix characteristics. So long as the mix stiffness moduli for the range in temperatures and a measure of its fatigue response are available, this can be accomplished.

When a fatigue expression like Eq 2 is available, the influence of different degrees of compaction (as measured by air-void content) and asphalt content on expected mix performance can be ascertained as well. This requires modifying Eq 2 by a functional relationship that includes the expression

$$\frac{V_B}{V_v + V_B} \qquad (4)$$

where V_B and V_v = volume of asphalt and air, respectively, in the mix. The 1981 thickness design manual of The Asphalt Institute [21] has incorporated this concept to adjust for expected voids in the asphalt concrete.

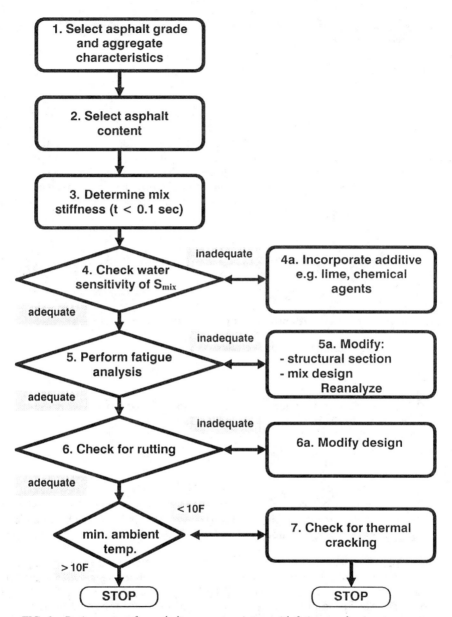

FIG. 6—*Design system for asphalt-aggregate mixture with fatigue as dominant concern.*

A simplified flow diagram illustrating the entire design process and emphasizing the fatigue mode of distress is shown in Fig. 6.

Rutting Subsystem

A general framework for estimating the potential for rutting is also illustrated in Fig. 5. As with fatigue, the stiffness characteristics of the mix at times of loading corresponding to moving traffic and for the temperatures expected at the site must be available.

While procedures to estimate the potential for rutting development are not as well defined as those for fatigue, one approach is to make use of the "layered-strain" predictive procedure [22]. In its most general form, relationships between permanent (plastic) strain, ε^p, applied stress, σ_{ij}, and load repetitions for each of the pavement components are required. At a particular number of load repetitions, the relationship can be stated as

$$\varepsilon^p = f(\sigma_{ij}) \tag{5}$$

For a particular layer, it is then possible to estimate the permanent deformation occurring in that layer. This is done by computing the permanent strain at a number of points within the layer, the number being sufficient to define the strain variation with depth. Permanent deformation is then determined by summing the products of the average permanent strains and the corresponding difference in depths between the locations at which the strains are determined, Fig. 7, that is,

$$\delta_i^p(x, y) = \sum_{i=1}^{n} (\varepsilon_i^p \Delta z_i) \tag{6}$$

where

$\delta_i^p(x, y)$ = rut depth in the ith position at point (x, y) in horizontal plane,

ε_i^p = average permanent strain at a depth of $\left(z_i + \dfrac{\Delta z_i}{2}\right)$, and

Δz_i = difference in depth.

To estimate permanent strain for a particular number of load repetitions in the vertical direction at a point in the pavement, use can be made of the expression

$$\varepsilon_z^p = R[\sigma_z - \tfrac{1}{2}(\sigma_x + \sigma_y)] \tag{7}$$

where

$\sigma_z, \sigma_x, \sigma_y$ = stresses in vertical and horizontal directions, respectively, Fig. 7; and
R = ratio of permanent (plastic) strain to stress that induces it at a specific number of stress applications and is obtained from laboratory repeated load tests.

At this time, it is not considered feasible to perform repeated load tests on the pavement materials for routine design purposes. Accordingly, a procedure developed by the Shell researchers [23], which is a simplification of this methodology, can be used for rut depth prediction. The Shell procedure makes use of the results of creep tests on asphalt concrete rather than repeated load tests.

Observations of the development of rut depths with load applications in test tracks provide data that, when suitably transformed, exhibit the same form as test results obtained from laboratory creep tests in uniaxial compression [23].

Results of the Shell procedure can be stated to include only the permanent deformation in the asphalt-bound layer

$$\Delta h_1 = C_M \cdot h_1 \frac{\sigma_{\text{ave}}}{S_{\text{mix}}} \tag{8}$$

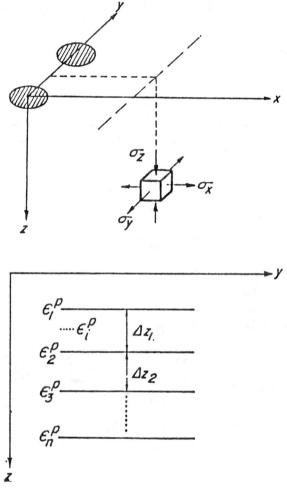

FIG. 7—*Schematic representation of pavement system used to estimate permanent deformation.*

where

- h_1 = thickness of the asphalt-bound layer,
- σ_{ave} = average stress in the asphalt-bound layer,
- S_{mix} = the value of the mix stiffness at $(S_{asp}) = [(S_{asp})]_v$ and determined as described in Ref 23, and
- C_M = correction factor for the so-called "dynamic effect" that takes account of differences between static (creep) and dynamic (rutting) behavior. This factor is dependent on the type of mix and has been found empirically to be in the range 1 to 2.[5]

---

[5] Reference 24 contains an example where the value of C_M was determined to be 1.5.

If the value of C_M were taken as equal to unity, and if it were desired to subdivide a thick asphalt-bound layer into a number of sublayers, the equation could be stated

$$\Delta h_1 = \sum_{i=1}^{n} \left[h_{1i} \times \frac{(\sigma_{ave})_{1i}}{(S_{mix})_{1i}} \right] \tag{9}$$

This procedure has been used, for example, in actual designs described by Finn et al. [25].

A possible framework for this process is included in Fig. 8. Available data suggest that rutting is most likely to occur at elevated temperature conditions. Accordingly, the steps that might be taken are as follows.

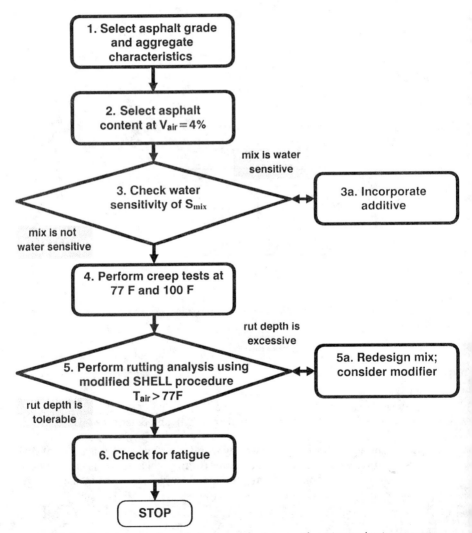

FIG. 8—*Design system for asphalt-aggregate mixture with rutting as dominant concern.*

(a) Perform creep tests on representative specimens to define S_{mix} as a function of time at 25°C (77°F), 40°C (100°F), and 60°C (140°F).

(b) Using the modified Shell procedure (as discussed herein) for air temperatures above 25°C (77°F), assess the potential for rutting.[6]

(c) If the analysis indicates that rutting is at an undesirable level for the expected conditions, the mix must be redesigned and the analysis repeated. The use of modifiers can also be considered.

Once a suitable mix has been established, the potential for other modes of distress can also be checked, for example, for fatigue as shown in Fig. 8.

Thermal Cracking Subsystem

This mode of distress is generally manifested by transverse cracks at the pavement surface. In qualitative terms, cracking is developed as follows. As the temperature at the pavement surface drops, there is a differential in pavement temperatures between the surface of the pavement and the interior since time is required for the "cold" to be conducted into the system. The surface of the pavement attempts to contract, but this contraction is restrained by the lower segments. Accordingly, stresses will develop due to the restraint of deformation. Initially, the stresses will be small since the stiffness of the mix is relatively low. However, as the temperature becomes colder, the tendency to deform is larger, but the lower segments of the pavement still prevent the deformation from occurring. Mix stiffness is also increased at the lower temperatures; the increased stiffness, coupled with the propensity for increased contraction, eventually leads to tensile stresses at the surface that exceed the breaking strength of the asphalt concrete and result, in turn, in surface cracking.

A number of approaches are possible to analyze the phenomenon [27,28], including treating the pavement as a:

(a) pseudo-elastic beam,
(b) approximate pseudo-elastic slab,
(c) viscoelastic slab, and
(d) approximate viscoelastic slab.

It might be argued that the beam assumption approximates conditions occurring at the edge of the pavement while the assumption of slab behavior provides conditions more severe than might be anticipated in the interior of the pavement [27].

Christison et al. [27] have demonstrated that the pseudo-elastic beam representation provides a "reasonable" estimate of the propensity for the asphalt pavements to develop cracks resulting from temperature changes in the low-temperature region. Recently, Anderson and Epps [29] have shown that this approach is useful in more moderate climates (for example, East Texas) to explain the transverse cracking occurring in "aged" asphalt pavements. This approach has been included as a design subsystem of the NCHRP 1-10B procedure, and the associated design framework is shown in Fig. 9.

---

[6] Suitable criteria will have to be established for the rutting analysis. For example, the magnitude of the allowable rut must be defined (less than 0.4 in. to mitigate hydroplaning in pavements with dense-graded mixes [26]) and the C factor of Eq 8 must be established to ensure that laboratory predictions correspond to rut depths obtained in practice.

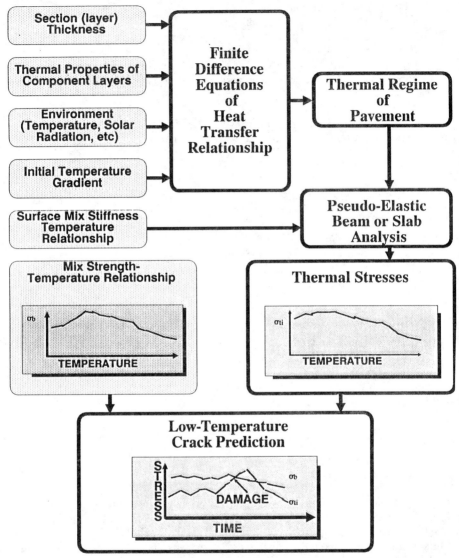

FIG. 9—*Low-temperature cracking.*

In this system mix, stiffness is required at a time of loading of 2 h (7200 s) in order to compute the thermal stress developed in the temperature interval, T_{i-1} to T_i.

The fracture strength of the mix is required in the low-temperature range to permit estimation of whether or not the predicted thermal stresses will lead to fracture. Like stiffness, this property is dependent on temperature. To minimize the testing required, use can be made of the results developed by Heukelom [9] that suggest that the tensile strength of a mix is dependent on the tensile strength of the asphalt contained in the mix, that is,

$$(\sigma_{\text{mix}})_b = M(\sigma_{\text{asp}})_b \qquad (10)$$

where

$(\sigma_{mix})_b$ = fracture (tensile) strength of mix;
$(\sigma_{asp})_b$ = tensile strength of asphalt contained in mix; and
M = mix factor, f(asphalt content, type, and grading of aggregate, void content).

This relationship can be used with limited tension test data (for example, as measured by the direct or indirect tension tests at one temperature) to define $(\sigma_{mix})_b$ over a range in temperatures.

Figure 10 illustrates the step-by-step sequence that might be followed. For cold climates, this response will govern the initial selection of mix characteristics as follows.

(a) Select *grade* of asphalt cement. Traffic conditions will influence selection of aggregate characteristics.
(b) Select asphalt content by available procedure (initial mix design).
(c) Define S_{mix} by creep tests in low-temperature range.
(d) Measure fracture strength of mix in the diametral test at $-17.8°C$ (0°F). Use the Shell approach, Eq 10, to modify results for other temperatures.
(e) Estimate propensity for low-temperature cracking using program COLD and weather data for a particular site.
(f) If mix is unsatisfactory, that is, cracking is indicated, revise mix design and repeat analysis. Asphalt modifiers may also be considered.

As for the other conditions, the pavement section must be known.

In the analysis to determine thermal stresses, temperatures are computed at a depth of 12.7 mm (0.5 in.) in the asphalt concrete surface. To compute these temperatures requires a knowledge of both the specific heat capacity and thermal conductivity of the mix. Reference *27* contains reasonable values for these parameters.

Raveling Subsystem

The three previous subsystems include a combination of materials testing and analyses of representative pavement structures to ascertain the adequacy of the mix. Estimation of the propensity for raveling has not, as yet, reached this stage. The framework illustrated in Fig. 5 does, however, provide general guidance.

An estimate of the type and amount of traffic is required. In addition, the environment, as defined by temperature and moisture conditions (including snow as well as water), will play a significant role. The system must also recognize the influence of the aging of the asphalt at and near the surface of the mix that will be more advanced than in the mix interior [*30*].

It is possible that a procedure like that developed by CALTRANS [*31*] may be utilized with mix performance criteria established on the basis of traffic and environmental conditions at the specific site.

Final Design Selection Considerations

Figure 1 shows that after all of the testing has been completed, a final decision must be made regarding materials selection, mix proportions, and any special provisions required for the specifications. Some of the factors that require evaluation include: materials availability, local experience, design policy, and project cost.

Users have very little control over the properties of asphalt that are to be used on a

60 ASPHALT CONCRETE MIX DESIGN

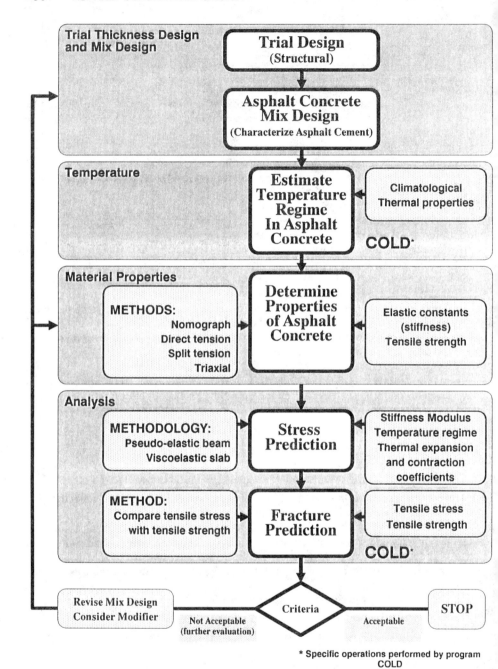

FIG. 10—*Low-temperature cracking subsystem.*

MONISMITH ET AL. ON A COMPREHENSIVE MIXTURE DESIGN SYSTEM   61

particular project as long as those asphalts comply with the prevailing specifications: for example, differences in temperature susceptibility or durability may exist within an area with multiple asphalt suppliers. Therefore in the final design, the specific asphalt (or modified asphalt) supplied by the contractor must be used to determine its influence on the mixture properties.

Local experience, design policy, and project cost may also impact on the final design. For example, local experience may dictate the design policy, that is, stabilized or unstabilized base, or more specifically, if local experience indicates the need for anti-stripping agents and laboratory tests fail to indicate such a requirement, local experience should prevail.

Construction Control Considerations

One of the unanswered questions relative to the role of asphalt concrete in pavement performance relates to the requirements for construction control and tolerance that can be applied in specifications. Two factors must be considered: (1) limitations of construction equipment to produce a uniform product, and (2) the effect of nonuniformity on performance.

With reliable performance prediction models, for example, based on the concepts described herein, it should be possible to establish the impact or consequences of various tolerance levels on the pavement life cycle through properly designed sensitivity analyses. Such evaluations permit the development of adjusted pay factors when materials are marginally "out of specifications" or indicate when the materials should be removed and replaced.

Specific Design Considerations

In the previous sections, a general framework for improved asphalt concrete mixture design has been presented. Not all of the steps have well-documented methodologies associated with them. In this section, some consideration will be presented on facets of the design process. Specifically, attention will be directed to specimen preparation for laboratory testing, to the use of the creep test as an adjunct to current design procedures, and to the development of criteria that will preclude extensive laboratory testing to evaluate such mix characteristics as fatigue response, and the propensity to develop permanent deformation under repetitive loading.

Laboratory Specimen Preparation

To ensure that specimens that are tested in the laboratory exhibit the same reponse as specimens of the same mixtures compacted in-situ require well documented comparisons of their stiffness and distress characteristics. This section briefly describes some of the type of data required to validate any laboratory compaction method.

The data presented herein provide a comparison of the creep behavior both in axial and shear loading of specimens compacted in the laboratory using the Triaxial Institute Kneading Compactor and specimens obtained by coring pavement sections containing the same materials. The sections from which the materials were obtained are located on Interstate 80 near Gold Run, CA. The two different mixtures tested were:

1. *Control Mix*—Asphalt concrete with dense graded aggregate and AR-4000 asphalt cement.

62   ASPHALT CONCRETE MIX DESIGN

2. *Carbon Black Reinforced Mix*—Asphalt concrete with the same dense graded aggregate and AR-4000 asphalt cement reinforced with carbon black (20% Microfil 8).

Specimens of these mixtures consisted of:

1. Cores.
2. Specimens mixed in the field and compacted in the laboratory using kneading compaction.
3. Specimens mixed in the laboratory and compacted in the laboratory using kneading compaction.

All of the specimens were tested in creep, both in axial and shear loading. Axial creep tests were performed at 20°C (68°F) and 38°C (100°F), while the shear creep tests were performed only at 38°C (100°F). An 0.2 MPa (30 psi) axial load was used for all axial creep tests conducted at 20°C (68°F), while a load of 0.15 MPa (20 psi) was used at 38°C (100°F) in order to preclude excessive deformation of the samples, as these same samples were then to be used for the shear creep tests.

The shear creep tests were performed on specimens obtained from the field and specimens mixed in the field and compacted in the laboratory. These specimens were cut to approximately 64 mm (2.5 in.) in height before testing as compared to the axial creep spec-

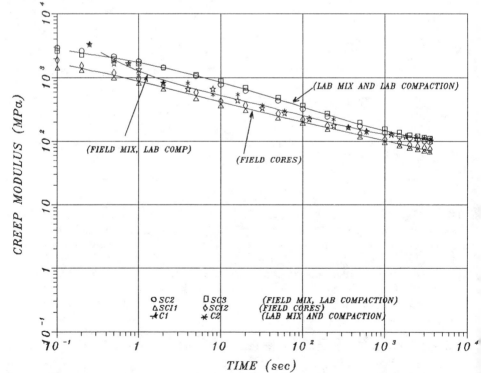

FIG. 11—*Comparison of axial creep response of laboratory-compacted specimens and field cores; 68°F, 30 psi applied stress; control mix.*

imens that were about 200 mm (8 in.) in height. Initially, 0.15 MPa (20 psi) and 0.07 MPa (10 psi) shear loads were used, but it was observed that the specimens failed within the first few seconds. The shear stress was then reduced to 0.038 MPa (5 psi) and all shear creep tests were carried out at this stress. In some field compacted specimens, especially the control specimens, failure in shear occurred at the layer interfaces. Interestingly, however, after the layer separation, the remaining halves of the specimens exhibited similar shear creep behavior to that obtained before layer separation. It should also be noted that the cores obtained from the field contained a fabric interlayer that might have influenced the axial creep behavior.[7] This fabric layer was removed by cutting the specimen before testing in shear creep.

Results of the tests are presented in Figs. 11 through 16, while Table 5 contains a summary of the specific gravity data for the specimens tested. Specific gravities of the field specimens containing carbon black are lower than those of the other specimens prepared in the laboratory. In general, these data indicate that specimens prepared by *kneading compaction* in the laboratory exhibit similar response, both in axial and shear loading in creep as specimens compacted by rolling in-situ.

The data also emphasize the importance of compaction on creep response. For the field

---

[7] The core specimens ranged from about 100 to 200 mm (4.0 to 4.6 in.) in height.

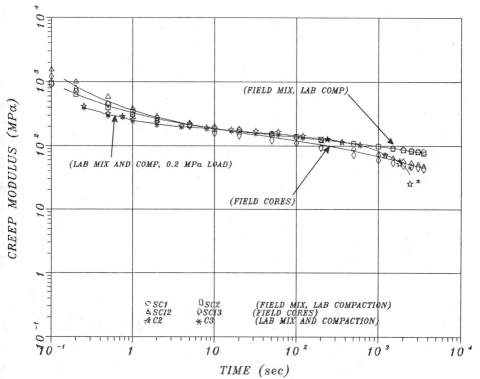

FIG. 12—*Comparison of axial creep response of laboratory-compacted specimens and field cores; 100°F, 20 psi applied stress; control mix.*

64 ASPHALT CONCRETE MIX DESIGN

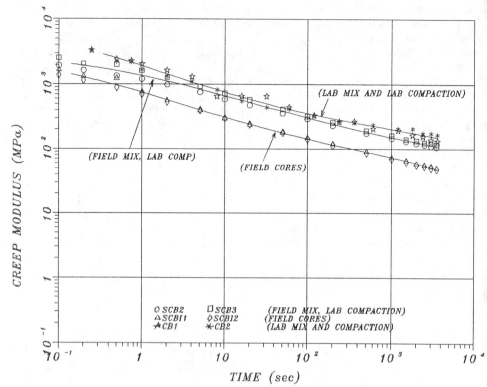

FIG. 13—*Comparison of axial creep response of laboratory-compacted specimens and field cores; 68°F, 30 psi applied stress; mixtures with carbon black reinforced asphalt.*

TABLE 5—*Specific gravities, laboratory specimens, and field cores for the Gold Run project.*

Field Cores		Field Mixed Laboratory Compacted Specimens		Laboratory Mixed Laboratory Compacted Specimens	
Specimen Designation	Specific Gravity	Specimen Designation	Specific Gravity	Specimen Designation	Specific Gravity
CONTROL MIX					
SCI 1	2.437	SC1	2.447	C1	2.430
SCI 2	2.430	SC2	2.432	C2	2.445
SCI 3	2.447	SC3	2.453	C3	2.450
MIX CONTAINING CARBON BLACK REINFORCED ASPHALT					
SCBI 1	2.405	SCB 1	2.470	CB1	2.436
SCBI 2	2.409	SCB 2	2.460	CB2	2.430
SCBI 3	2.403	SCB 3	2.453	CB3	2.456

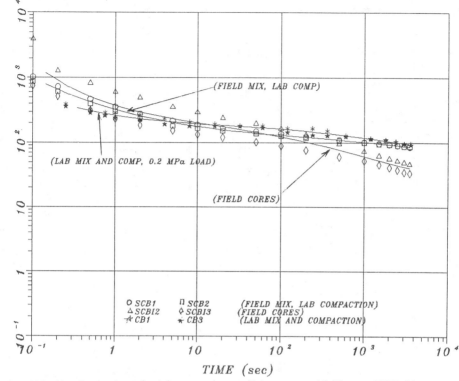

FIG. 14—*Comparison of axial creep response of laboratory and field cores; 100°F, 20 psi applied stress; mixtures with carbon black reinforced asphalt.*

cores containing the carbon black reinforced asphalt, the densities (specific gravities) were lower and the creep response (Figs. 13, 14, and 16) correspondingly was at a lower level.

Comparison of the data for the specimens containing the control asphalt and the carbon black modified material indicate an interesting phenomenon. While not shown here, the stabilimeter values for both mixes are essentially the same. Similarly, the response in axial creep, when specimens are compared at comparable densities, is also essentially the same. On the other hand, the behavior of the specimens containing the carbon black modified asphalt in shear creep is improved over that for the specimens containing the control asphalt, at least as measured by the creep modulus in shear computed in this study.

Use of the Creep Test in the Mix Design Process

As a part of current mix design methodology, it may be useful to consider the creep test as a part of the mix design process for mixtures containing modified asphalts.

One of the questions that arises for mixes containing these materials is associated with the quantity of modifier to use. Figure 17 presents the results of creep tests at 43°C (110°F) on specimens containing a granite aggregate and an AR-2000 asphalt cement modified with varying proportions of carbon black[8] [*32*]. It will be noted that the specimens fail in creep

---

[8] In the form of Microfil 8 supplied by the Cabot Corporation.

66 ASPHALT CONCRETE MIX DESIGN

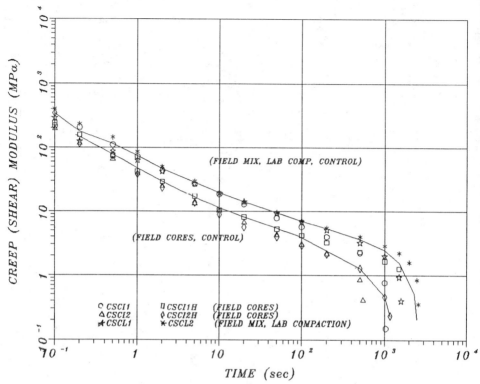

FIG. 15—*Comparison of shear creep response of laboratory-compacted specimens and field cores; 100°F, 5 psi applied shear stress; control mix.*

at this temperature with treatment levels of 5 and 10% microfiller (by weight of asphalt). On the other hand, the response of specimens containing 15 to 20% of the modifier suggests that treatment levels in this range will produce mixes that will perform at least as well as specimens containing an AR-8000 cement from the same source as the carbon black modified material. Interestingly, this level of modification corresponds to that arrived at by field experience with these materials. It should be noted that this use was to attempt to mitigate low-temperature cracking and at the same time reduce the propensity for rutting at high temperatures [*32*].

Mix Design and Testing Considerations to Simplify Laboratory Evaluations

In the design subsystems to consider the various modes of distress, especially for fatigue, rutting, and thermal cracking, it may be desirable to develop criteria that would minimize the amount of testing required. For example, it may be difficult under many circumstances to justify an extensive fatigue testing program to evaluate the suitability of specific mixtures for the project under consideration. On the other hand, it is possible that a relatively simple screening test may provide an indication of such performance. The LCPC of France has suggested such an approach in which a single axial tensile loading test is utilized [*15*]. A test sequence for a specimen of the mix under investigation is followed using that shown in Table 6. The tensile strain corresponding to 10^6 applications is estimated from a

MONISMITH ET AL. ON A COMPREHENSIVE MIXTURE DESIGN SYSTEM 67

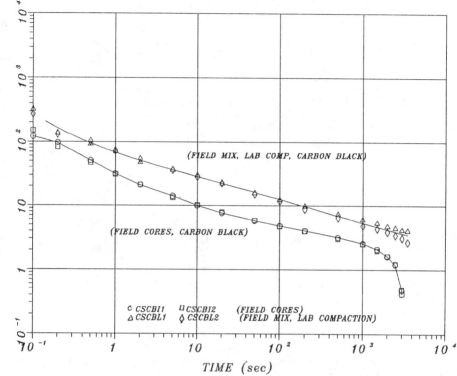

FIG. 16—*Comparison of shear creep response of laboratory-compacted specimens and field cores; 100°F, 5 psi applied shear stress; mixtures with carbon black reinforced asphalt.*

TABLE 6—*Test conditions for simplified procedure for mix evaluation under tensile loading.[a]*

Order	Temperature, °C	Selected Nominal Speed	Strain	Recovery Time, min
1	10	$1/\sqrt{60}$ mm/s	10^{-4}	5
2	10	1 mm/min	10^{-4}	15
3	10	$1/\sqrt{60}$ mm/min	10^{-4}	...
4	20	$1/\sqrt{60}$ mm/s	$2 \cdot 10^{-4}$	5
5	20	1 mm/min	$2 \cdot 10^{-4}$	15
6	20	$1/\sqrt{60}$ mm/min	$2 \cdot 10^{-4}$	...
7	0	$1/\sqrt{60}$ mm/s	10^{-4}	5
8	0	1 mm/min	10^{-4}	15
9	0	$1/\sqrt{60}$ mm/min	10^{-4}	15
10	0	$1/\sqrt{60}$ mm/min	break	...

[a] After ref (*15*).

68 ASPHALT CONCRETE MIX DESIGN

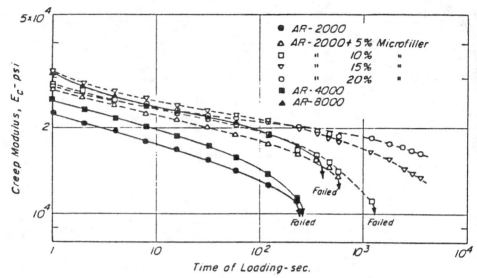

FIG. 17—Creep modulus versus time of loading, 110°F; granite aggregate.

regression equation using parameters established for the particular mix during the test sequence. Data that the French have presented indicate excellent correlation between the estimated strain and that measured in controlled strain fatigue tests at 10°C conducted at a frequency of 25 Hz.

With the data obtained from the test program, an elastic quality indicator (IQE) is calculated. This IQE is the thickness of asphalt concrete required to sustain 10^6 repetitions (using estimated strain at $N = 10^6$ from tension test sequence) of a 130 kN axle load when the modulus of the subgrade is 100 MPa and the stiffness of the mix is determined from the test data at 10°C and at a time of loading of 0.02 s. A multilayer elastic program (ALIZE) is used to arrive at the thickness of asphalt concrete in the analysis process. In this estimate, the IQE decreases as the quality of the asphalt increases.

Table 7 has been prepared to indicate some considerations that will influence the final design, including examples of criteria that may be of assistance to minimize: (1) testing to define specific mix parameters, and (2) analyses to estimate the propensity for specific distress conditions.

In this table, the endeavors of a number of investigators have been incorporated. It will be noted, for example, that the criteria developed by the LCPC for fatigue and referred to previously are briefly noted. A number of different investigators have suggested criteria to be associated with the creep test to mitigate rutting. Unfortunately, the limiting values shown in Table 7 are associated with creep tests that differ in the method of specimen preparation, specimen height, and end conditions, etc. Nevertheless, the concept of a limiting value of creep modulus appears to be a useful adjunct to assist the mix designer in assessing the efficacy of a mix for conditions of traffic beyond those for which some of the conventional procedures are applicable.

Summary

In this paper, an approach has been presented to the development of an asphalt aggregate mixture analysis system (AAMAS) that incorporates the latest state of the technology with

Mix Property	Fatigue	Rutting	Thermal Cracking (Fracture)	Raveling
Stiffness	mix stiffness, S_{mix} at $t = 0.01$s, 20°C (68°F) 1. Thin asphalt-bound layer 700 to 1400 MPa (100 000 to 200 000 psi) 2. Thick asphalt-bound layer >3500 MPa (500 000 psi) mix stiffness, S_{mix} at $t = 300$ s, 0°C, used by LCPC as one of the parameters to estimate fatigue response [15]	mix stiffness, S_{mix} at 40°C (104°F) 1. Ref 33: $S_{mix} \geq 80$ MPa (12 000 psi) at $t = 100$ min and $\sigma_0 = 0.2$ MPa (30 psi) 2. Ref 34: $S_{mix} \geq 50$ to 65 MPa (7500 to 10 000 psi) at $t = 60$ min and $\sigma_0 = 0.1$ MPa 3. Ref 25: $S_{mix} \geq 135$ MPa (20 000 psi) at $t = 60$ min and $\sigma_0 = 0.2$ MPa (30 psi)	mix stiffness, S_{mix} at lowest temperature expected in-situ 1. Ref 35: $S_{mix} < 20$ GPa (3×10^6 psi) at $t = 30$ min	1. asphalt stiffness, S_{asp}, at $t = 10^{-3}$ s less than some value for low road temperature. 2. strain at break for mixture $(\epsilon_b)_{mix}$ at $t = 0.01$ s larger than some value (for example, 10^{-1}) for expected low pavement temperature [36].
Durability (changes)	1. Thin asphalt-bound layer increased stiffness may lead to reduced fatigue life 2. Thick asphalt bound layer change in stiffness less significant	rutting resistance improved with stiffness increase.	increase in stiffness increases propensity for thermal cracking.	increase in asphalt stiffness will lead to increased potential for surface abrasion at low pavement temperatures [30,36]. For example, for ADT > 3000, loss of matrix higher as η at 140°F for recovered asphalt is higher [30].
Fracture	strain at break in direct tension used by LCPC as one of the parameters to estimate fatigue response [15]		improved fracture strength will increase resistance of mix to thermal cracking.	
Water sensitivity (stiffness reduction)	reduced stiffness will lead to reduction in fatigue life in heavy duty asphalt concrete sections.	reduced stiffness will lead to increased rutting.	will lead to reduction in fracture strength of mix as well as stiffness. Propensity for thermal cracking will be increased.	

[a] This table should be considered as an initial attempt; additional criteria can be included as such information is developed.

due consideration to its application to present day practice. Such an approach would appear timely because of the increasing lack of confidence in present-day test methods that have been found to be inadequate to produce mix designs for situations other than routine applications.

Research, like that summarized herein, has provided requisite analytical methodology based on improved considerations of pavement response to traffic and environmental influences together with test equipment and procedures that should be seriously considered for more widespread use.

Whatever system that is finally developed should have general applicability to the widest ranges of environmental and traffic loading conditions such that the engineer can formulate a mix design that will achieve specific performance objectives with a high degree of confidence.

It is the considered opinion of the authors that an asphalt concrete mixture design system incorporating the principles and methodologies presented and discussed in this paper will accomplish this objective.

Acknowledgments

The authors express their thanks to Dr. Jorge Sousa and Mr. A. Tayebali for the axial and shear creep test data included in this paper to illustrate a methodology for comparing the stress versus strain characteristics of laboratory-compacted and field-compacted specimens and to the staff of the Transportation Laboratory of CALTRANS for supplying the materials and cores to make this comparison possible.

Ms. Phyllis De Fabio typed the manuscript.

References

[1] Vallerga, B. A., "The Effects of Aggregate Characteristics on the Stability of Asphaltic Paving Mixtures," paper presented at the 41st Annual Convention of the National Sand and Gravel Association, Los Angeles, 1957.
[2] Salt, G. F., "Research on Skid Resistance at the Transportation Research Laboratory (1927–77)" *Skidding Accident Characteristics,* Transportation Research Record 120, Transportation Research Board, Washington, DC, 1976.
[3] Schmidt, R. J. and Graf, P. E., "The Effect of Water on the Resilient Modulus of Asphalt Treated Mixes," *Proceedings,* Association of Asphalt Paving Technologists, 1972.
[4] Anderson, D. A. and Goetz, W. H., "Mechanical Behavior and Reinforcement of Mineral Filler—Asphalt Mixtures," *Proceedings,* Association of Asphalt Paving Technologists, 1973.
[5] Dukatz, E. L. and Anderson, D. A., "The Effect of Various Fillers on the Mechanical Behavior of Asphalt and Asphaltic Concrete," *Proceedings,* Association of Asphalt Paving Technologists, 1980.
[6] Anderson, D. A. and Tarris, J. P., "Characterization and Specification of Baghouse Fines," *Proceedings,* Vol. 52, Association of Asphalt Paving Technologists, 1983.
[7] Anderson, D. A., Tarris, J. P., and Brock, J. D., "Dust Collector Fines and Their Influence on Mixture Design," *Proceedings,* Association of Asphalt Paving Technologists, 1982.
[8] "Strategic Highway Research Program—Research Plan," Transportation Research Board, Washington, DC, May 1986.
[9] Heukelom, W., "Observations on the Rheology and Fracture of Bitumen and Asphalt Mixes," *Proceedings,* Association of Asphalt Paving Technologists, 1966.
[10] Heukelom, W., "Viscosities of Dispersions as Governed by Concentration and Rate of Shear," *Proceedings,* Association of Asphalt Paving Technologists, 1971.
[11] Heukelom, W., "An Improved Method of Characterizing Asphaltic Bitumens with the Aid of Their Mechanical Properties," *Proceedings,* Association of Asphalt Paving Technologists, 1973.
[12] Epps, J. A., "Asphalt Pavement Modifiers," *Civil Engineering,* April 1986, p. 56.
[13] "Method of Test for Centrifuge Kerosene Equivalent and Approximate Bitumen Ratio," Test Method 303, CALTRANS, 1978.

[14] Nijboer, L. W., *Plasticity as a Factor in the Design of Dense Bituminous Carpets*, Elsevier Publishing Company, Amsterdam, 1948.
[15] Bonnot, J., "Asphalt Aggregate Mixtures," Transportation Research Record No. 1096, Transportation Research Board, Washington, DC, 1986, pp. 42–51.
[16] Lottman, R. P., "Predicting Moisture—Induced Damage to Asphaltic Concrete—Field Evaluation," NCHRP Final Report 246, Transportation Research Board, Washington, DC, May 1982.
[17] Monismith, C. L., Epps, J. A., and Finn, F. N., "Improved Asphalt Mix Design," *Proceedings, Association of Asphalt Paving Technologists*, Vol. 54, 1985.
[18] Ahlborn, G., "ELSYM5, Computer Program for Determining Stresses and Deformation in Five Layer Systems," University of California, Berkeley, 1972.
[19] DeJong, D. L., Peutz, M. G. F., and Korwagen, A. R., *Computer Program BISAR, Layered Systems Under Normal and Tangential Surface Loads*, Koninklijke Shell—Laboratorium, Amsterdam, External Report AMXR 0006.73, 1973.
[20] *AASHTO Guide for the Design of Pavement Structures*, American Association of State Highway and Transportation Officials, Washington, DC, 1986.
[21] *Thickness Design—Asphalt Pavements for Highways and Streets*, Manual Series No. 1, The Asphalt Institute, College Park, MD, 1981.
[22] Monismith, C. L., Inkabi, K., Freeme, C. R., and McLean, D. B., "A Subsystem to Predict Rutting in Asphalt Concrete Paving Structures," *Proceedings*, Vol. 1, Fourth International Conference on the Structural Design of Asphalt Pavements, University of Michigan, Ann Arbor, 1977.
[23] van de Loo, P. J., "A Practical Approach to the Prediction of Rutting in Asphalt Pavements—The Shell Method," Transportation Research Record No. 616, Transportation Research Board, Washington, DC, 1976.
[24] Monismith, C. L., Finn, F. N., Ahlborn, G., and Markevich, N., "A General Analytically Based Approach to the Design of Asphalt Concrete Pavements," *Proceedings*, Vol. 1, Sixth International Conference on the Structural Design of Asphalt Pavements, University of Michigan, Ann Arbor, July 1987.
[25] Finn, F. N., Monismith, C. L., and Markevich, N. F., "Pavement Performance and Asphalt Concrete Mix Design," *Proceedings, Association of Asphalt Paving Technologists*, Vol. 52, 1983.
[26] Monismith, C. L. and D. B. McLean, *Design Considerations for Asphalt Pavements*, Report No. TE 71-8, University of California, Berkeley, 1971.
[27] Christison, J. T., Murray, D. W., and Anderson, K. O., "Stress Prediction and Low Temperature Fracture Susceptibility of Asphalt Concrete Pavements," *Proceedings, Association of Asphalt Paving Technologists*, Vol. 41, 1972.
[28] Shahin, M. L., "Design System for Minimizing Asphalt Concrete Thermal Cracking," *Proceedings*, Vol. 1, Fourth International Conference on Structural Design of Asphalt Pavements, University of Michigan, Ann Arbor, 1977.
[29] Anderson, K. O. and Epps, J. A., "Asphalt Concrete Factors Related to Pavement Cracking in West Texas," *Proceedings, Association of Asphalt Paving Technologists*, Vol. 52, 1983.
[30] Vallerga, B. A. and Halstead, W. J., "Effects of Field Aging on Fundamental Properties of Paving Asphalts," *Highway Research Record No. 361*, Highway Research Board, Washington DC, 1971, pp. 71–92.
[31] "Method of Test for Surface Abrasion of Compacted Bituminous Mixtures," Test Method 360, State of California Department of Transportation, 1978.
[32] Yao, Z. and Monismith, C. L., "Behavior of Asphalt Mixes with Carbon Black Reinforcement," *Proceedings, The Association of Asphalt Paving Technologists*, Vol. 55, 1986, pp. 564–585.
[33] Viljoen, A. W. and Meadows, K., *The Creep Test—A Mix Design Tool to Rank Asphalt Mixes in Terms of Their Resistance to Permanent Deformation Under Heavy Traffic*, National Institute for Transport and Road Research, Pretoria, South Africa, Technical Note: TP/36/81, May 1981.
[34] Kronfuss, F., Krzemien, R., Nievelt, G., and Putz, P., *Verformungsfestigkeit von Asphalten Ermittlung im Kriechtest*, Bundesministerium fur Bauten und Technik, Strassenforschung, Heft 240, Wien, Austria, 1984.
[35] Gaw, W. J., Burgess, R. A., and Young, F. D., "Ste Anne Test Road—Road Performance After Five Years and Laboratory Predictions of Low Temperature Performance," *Proceedings*, Canadian Technical Asphalt Association, Vol. 19, 1974.
[36] Dormon, G. M., "Some Observations on the Properties of Bitumen and Their Relation to Performance in Practice and Specifications," *Proceedings*, First Conference on Asphalt Pavements for Southern Africa, Durban, South Africa, July–Aug. 1969.

Byron E. Ruth,[1] Mang Tia,[1] and Kwasi Badu-Tweneboah[1]

The Role of Asphalts in Rational Mix Design and Pavement Performance

REFERENCE: Ruth, B. E., Tia, M., and Badu-Tweneboah, K., "**The Role of Asphalts in Rational Mix Design and Pavement Performance,**" *Asphalt Concrete Mix Design: Development of More Rational Approaches, ASTM STP 1041*, W. Gartner, Jr., Ed., American Society for Testing and Materials, Philadelphia, 1989, pp. 72–85.

ABSTRACT: Conceptually, a rational mix design procedure should provide test parameters for input into mechanistic pavement design methods and analytical procedures to predict high and low-temperature pavement performance. It is suggested that high-temperature evaluation be performed to design asphalt concrete mixtures using a gyratory testing machine (GTM) to simulate (1) field compaction at conventional mix temperatures, and (2) traffic densification at 60°C. This would provide a better indication of mixture behavior and sensitivity to changes in aggregate blend and asphalt content than current mix design methods. The major advantage of this approach is that the changes in the shear response of the mix can be evaluated easily over the range from as-compacted to highly densified. Conventional mix design procedures are essentially one-point tests that do not provide information rate of densification or changes in stability when the mix density changes.

Tentative recommendations are presented for quality control and acceptance testing using gyratory testing of plant-produced hot mix. Advantages of this procedure are its sensitivity to combined changes in gradation and asphalt content, simplicity, and potential for more intensive testing of produced hot mix.

The low temperature (<25°C) properties and behavior of asphalt concrete paving mixtures are controlled by the rheological properties of the asphalt concrete. Therefore, it is essential that specifications for asphalts or mixtures or both be established to minimize the potential for excessive age hardening and pavement cracking. Research investigations have found cracking to be related to thermal contraction, thermal gradients that produce pavement rippling, vehicular loads, and the viscosity of the asphalt binder during cooling of the pavement. A stress analysis is presented to illustrate the combined effect of asphalt concrete moduli, underlying pavement support, and heavy wheel loading on the maximum pavement stresses developed at designated low temperatures.

A brief discussion on the selection of asphalts and polymer modification is presented to illustrate the major deficiencies in current test methods and specifications. Most asphalt specifications require only tests at 25°C and higher that neglect behavior at low temperatures. Low-temperature viscosity tests should be incorporated in all asphalt specifications. A comprehensive model asphalt specification is presented that includes viscosity tests at 15 and 25°C. An example is given to demonstrate how the viscosity-temperature relationships enhance our ability to interpret asphalt, polymer modified asphalt, and mix behavior at low temperature.

KEY WORDS: mix design, asphalt specifications, gyratory testing machine, viscosity, low temperatures, stress analysis, pavement performance, thermal contraction, thermal rippling, polymers, asphalt concrete

[1] Professors and doctoral candidate, respectively, Department of Civil Engineering, University of Florida, Gainesville, FL 32611.

Historically, the interaction between mix design and pavement performance has been limited to empirical correlations or visual observations of the behavior of asphalt mixtures at high ambient temperatures (generally above 25°C). The primary concern has been whether or not the paving mixture provided adequate stability to minimize the potential for shoving and rutting. Efforts to improve mixture properties and quality have generally involved one or more of the following changes:

1. An increase in compactive effort (for example, 50 to 75 blow Marshall).
2. An increase in stability requirements for structural mixtures from 4.45 kN (1000 lb) to 8.06 kN (1800 lb) minimum stability.
3. Reduction in maximum allowable flow (Marshall test).
4. An increase in specified minimum percent compaction required for paving operations.
5. The adoption of quality control and quality assurance specifications for asphalt pavement construction (often including a penalty clause for payment).

Although many of these changes may be beneficial for improving the high-temperature performance of asphalt paving mixtures, the key to good mix design depends on the selection of suitable aggregates and job mix aggregate gradation that will yield adequate VMA, air void content, and stability of the in-place pavement layer. A well-designed mix should provide suitable as-compacted and long-term performance without excessive densification, loss in stability, or abnormal age hardening of the asphalt binder that influences low-temperature performance and susceptibility to pavement cracking.

The low-temperature (generally below 25°C) behavior of asphalt concrete mixtures, assuming adequate structural capacity of underlying pavement and foundation materials, depends almost exclusively upon the elastic and viscous properties of the asphalt binder. Asphalts (bitumens) become extremely brittle at sufficiently low temperatures. Their creep deformation even under high stresses is almost nonexistent resulting in a very brittle material with extremely low tolerance to strain and to energy induced by heavy vehicle loadings and thermal effects (contraction or rippling or both from temperature differentials). Asphalts in a brittle state (near or above T_g) exhibit a high modulus or stiffness that is beneficial in reducing stresses in underlying layers. However, low-temperature cracking potential is high when subjected to large thermal gradients. In general, the viscosity of the asphalt is directly correlated to the properties of asphalt concrete paving mixtures.

Current asphalt and asphalt mixture specifications are for the most part totally inadequate for rational assessment of properties throughout the range of in-service pavement temperatures. It may be rationalized and demonstrated by suitable testing techniques that any test conducted at one temperature does not define the behavior of these materials at other temperatures. Numerous tests that are performed do not yield parameters that are usable in the prediction of pavement performance or the modeling of pavement response to induced loads. Parameters such as penetration ratio (Penetration 4°C/Penetration 25°C) and viscosity ratio (V_{60TFOT}/V_{60ORIG}) give only an indication of temperature susceptibility and early asphalt hardening, respectively. Also, the Penetration/Viscosity ratio is being used to some extent to select suitable asphalts for the design of flexible pavements.

Another major deficiency relating to low-temperature pavement performance is our inability to predict the age hardening of different asphalts. Current test methods do not provide quantitative data on changes in asphalt properties at the asphalt hot-mix plant or on the roadway. The major variables that influence the degree of hardening are:

1. Asphalt characteristics that are related to the origin of the crude oil and refining methods used to obtain the asphalt cement.
2. Asphalt plant hot-mix temperatures and the interaction of aggregate moisture with type of mixing (batch plant, drum mixer, coater unit, type of fuel, etc.).
3. Air void content (permeability) of the compacted asphalt concrete pavement and the degree of sealing of its surface by pneumatic rolling, traffic, or application of dense overlay, sealcoat, etc.

In certain situations, aggregate mineralogical and pore structure contribute to changes in asphalt and mix properties. Factors such as poor asphalt-to-aggregate bond, stripping, selective adsorption, or retained moisture (particularly in highly absorptive aggregates) can individually or in combination result in early distress of asphalt pavements.

It is obvious that it may be difficult to solve all of the aforementioned problems. However, it is possible to develop an improved understanding of the relationship between asphalt-mixture-pavement behavior that can eventually lead to improved (more rational or mechanistic) specifications, analysis techniques, and pavement design procedures.

Mix Design Concepts: High-Temperature Pavement Performance

The components of good mix design should incorporate:

1. Guidelines (specifications) for gradation of the aggregate blend and for selection of aggregate types for blending to meet gradation requirements.
2. Laboratory-mix temperatures comparable to specifications for plant-mix temperatures.
3. Laboratory compaction that simulates the maximum field-compacted densities at suitable air void contents attained using conventional equipment and good rolling procedures.
4. Stability tests on compacted specimens at different asphalt contents to determine their resistance to consolidation rutting and plastic deformation.
5. Traffic densification simulation tests to evaluate the long-term (for example, two to five years) effects on resistance to consolidation rutting and plastic deformation.

The purpose of mix design is to provide resistance to rutting without adversely affecting the durability of the mix. Most cases of severe rutting or shoving occur almost immediately or within two years after construction. Current design methods (for example, Marshall and Hveem methods) may yield good results but their major deficiency is the lack of test results throughout the density spectrum (as-compacted to traffic-densified). A mix design procedure similar to that proposed using the gyratory testing machine is a logical way to approach the high-temperature pavement performance problem.

The gyratory testing machine (GTM) has the inherent characteristics necessary for mix design following field-simulation procedures. Table 1 presents data on different mixtures that were compacted at 8 and 12 gyratory revolutions at 146°C for comparison to field-compacted mixtures. Twelve revolutions produced almost exactly the same density and air void content levels as obtained with field compaction. Kneading compaction at 300 psi foot pressure is intended to simulate field-compacted densities. The Hveem stability values for gyratory-compacted specimens (Table 1) are obviously lower than those for kneading-compacted specimens. Open-graded mixtures gave almost the same Hveem stability values for cores and gyratory-compacted specimens. Greater variations between field and gyratory stability values occurred with the dense-graded mixtures.

TABLE 1—*Gyratory field compaction simulation results.*

Mix Characteristics	Extraction Results		Field Compaction as a Percent of Gyratory[a]		Percent Air Voids at 12 Revolutions	Hveem Stability				
	Asphalt Content, %	Mineral Filler, %	8 Passes/ Revolutions	12 Passes/ Revolutions		Cores		Lab, 12 Revolutions	Kneading, 300 psi[f]
						0 months	9 months		
DENSE GRADED										
(8) LS + MF	5.4	5.5	101.0	101.0	5.4	30	33.0	27.2	34.0[g]
(11) G + RS + MF	5.2	5.9	100.0	100.0	4.2	25.4	25.8	32.4	30.5
(14) LS + RS + MF	5.4	6.0	99.7	99.6	4.3	19.7	24.4	31.7	35.0
(17) SS + RS + MF	8.0	5.1	99.7	99.6	6.3	23.7	22.4	...	...
(19) LS + SS + MF	10.4	9.0[e]	99.6	99.4	8.8	20.0	18.2	35.0	36.0
Mean, $\bar{x}$			100.0	99.9					
OPEN GRADED (DEFICIENT MF)										
(6) LS[b]	6.8	2.6	102.1	101.6	(4.8)	18.8	23.8	22.4	...
(7) LS[b]	6.3	2.9	102.4	98.9	(4.8)	26.4	27.0	24.0	34.0[g]
(9) G + RS[b]	7.2	1.8	99.9	99.8	(4.7)	25.0	28.5	22.8	24.5
(10) G + RS[b]	7.8	1.6	100.1	99.3	(4.7)	24.4	26.2	...	...
(12) LS + RS	6.8	1.8	100.6	100.0	(4.9)	25.6	28.9	27.1	30.0
(13) LS + RS	7.3	1.8	99.7	99.8	(4.9)	23.2	31.0	24.8	...
(15) S + RS	8.2	2.4	102.7	102.0	(6.7)	29.7	32.9	33.0	39.0[g]
(16) S + RS[c]	10.6	1.5	107.3[d]	107.3[d]	(6.7)	25.8	23.9	21.0	...
Mean, $\bar{x}$ (omit No. 16)			101.1	100.2					

NOTES—

Limestone = LS, mineral filler = MF, crushed gravel = G, river sand = RS, slag = S, slag sand = SS, and ( ) = approximate values.

[a] Gyratory Model 4-C: 3° angle, 689.5 kPa (100 psi) ram, 103 kPa (15 psi) air roller. Percent compaction based on number of roller coverages (passes) being equivalent to the number of gyratory revolutions. Aggregate and asphalt heated to 146 ± 3°C for mix preparation. Mix temperature during compaction was in the range of 130 to 135°C.

[b] Aggregate gradation adjusted to conform to that placed in the field.

[c] Overasphalted and excessive moisture in mix.

[d] Omitted from mean percent compaction.

[e] Design MF content was 7.0%.

[f] Simulation of field-compacted densities.

[g] Hveem stability values substantially greater than obtained on cores and gyratory specimens.

TABLE 2—*Gyratory traffic densification simulation results.*

Mix Characteristics	Hveem Stability after Densification[a]	Air Void Content Laboratory Samples, %
DENSE GRADED		
(8) LS + MF	32.8	1.5
(11) G + RS + MF	27.5	1.5
(14) LS + RS + MF	31.8	1.7
(17) S + RS + MF	53.1	2.9
(19) LS + SS + MF	50.3	2.9
OPEN GRADED (DEFICIENT MF)		
(6) LS ...	...	...
(7) LS 33.2	0.8	...
(9) G + RS	...	...
(10) G + RS	32.8	1.2
(12) LS + RS	...	...
(13) LS + RS	35.7	1.3
(15) S + RS	...	...
(16) S + RS	51.3	2.8

[a] Gyratory compacted specimens: 3° angle, 689.5 kPa (100 psi) ram, 103 kPa (15 psi) air roller, and 12 revolutions. Densified at 60°C to simulate traffic: 2° angle, 689.5 kPa (100 psi) ram, 138 kPa (20 psi) air roller, and 200 revolutions.
See Refs *1* and *2* for detailed information.

Traffic densification simulation was performed at 60°C using the GTM. Table 2 gives the Hveem stability and air void content values corresponding to a gyratory shear, G_s value, of 27.0 after 200 revolutions of densification. During the densification process, the G_s value (stability) will generally increase and then decrease. The rate (number of revolutions) at which this occurs is dependent upon aggregate blend and asphalt content. Aggregate blends containing slag or hard angular aggregates (for example, trap rock) may continue to densify without any significant change in the G_s value. These effects on stability are probably related to the VMA during densification and the angularity and surface texture of the aggregate. The Hveem stability values range from 27.5 to 35.7 for the different mixtures except those containing slag that were in the 50s. The high stability values for the slag mixtures may be attributed to aggregate particle interlock at strain levels that are lower in the Hveem stability test than produced by the GTM. This gyratory design procedure may require slight alterations in testing procedures (for example, temperature, etc.) to conform to regional construction and environmental conditions.

Suggestions for Design and Quality Control of Mixtures

It is envisioned that gyratory compaction and densification procedures using variable stress and strain conditions (air roller or spring with suitable spring constant) can be developed to duplicate achievable field-compacted density levels and to simulate the traffic densification process. The primary advantages of a gyratory design procedure are the following:

1. Shear resistance of the mixture can be assessed throughout the range in density from as-compacted to traffic densified.

2. The combined effects of aggregate gradation, particle shape and texture and asphalt content can be evaluated without direct dependency upon air void content and VMA. However, a knowledge of air void contents and VMA can be useful for modification of aggregate blend for improvement of mix properties when gyratory shear results are not satisfactory.
3. If desired, the design mix can be evaluated to simulate acceptable field production variations in gradation and asphalt, particularly a combined increase in mineral filler and asphalt content.

Quality control or acceptance testing of hot mix at the plant, or both, can be improved and expedited using gyratory compaction with increased revolutions to evaluate gyratory shear response. Shortly after the start of production, gyratory tests and extraction tests can be performed in unison. If the extraction tests verify conformance to the job mix formula, the gyratory shear response band (variation between samples) can be established for evaluation of subsequent hot-mix production for a paving project. A change in gyratory shear response from the established reference would indicate a significant change in mixture characteristics that could be evaluated by extraction tests.

Exploratory investigations performed in the past indicate this gyratory testing method is very sensitive to change in asphalt content and aggregate gradation for many different types of aggregates. However, highly angular and open-graded aggregates may be insensitive to changes in asphalt content unless it is grossly excessive.

The major advantage of a suitable gyratory testing procedure for plant control and acceptance testing is its reliability in detecting mix quality during production that may be related to rutting resistance. Another key factor is the simplicity of the test. Testing personnel could perform and evaluate test results in less than 10 min. Consequently, production could be monitored on a more frequent basis to minimize the potential of having a substantial tonnage of defective hot mix produced before the problem is detected.

Low-Temperature Performance of Asphalt Pavements

Research investigations conducted since 1972 have continually verified the importance of asphalt consistency (for example, penetration and viscosity) for the evaluation of mixture properties and pavement behavior at low temperatures. The concept that low-temperature thermal cracking occurs only at high latitudes or in extremely cold climates is false since the viscosity of the asphalt binder dictates the actual range of critical low temperature. Furthermore, low-temperature cracking of pavements depends upon traffic load induced stresses that are affected by the structural response of the pavement. During cooling and contraction of the asphalt pavement, tensile stresses are developed when the rate of cooling and viscosity of the asphalt are sufficiently high to prevent creep relaxation of the induced strains. This problem is further accentuated by large temperature differentials between the surface and bottom of the asphalt concrete that induces thermal "rippling."

The rippling effect, which is similar to curling of concrete pavements, produces lift-off of the asphalt concrete from the base course in a low-amplitude sinusoidal form as illustrated in Fig. 1. As pavement cooling progresses, the asphalt viscosity and mix stiffness increases thereby reducing the creep strain rate that increases the contraction and rippling (bending) stresses. When rippling is initiated (Fig. 1a) the wave form has a short wave length and low amplitude. Heavy traffic during this stage will increase stresses, produce dynamic creep, and result in a reduction or elimination of the low-amplitude lift-off. Further cooling of the pavement in the temperature range approaching the T_g of the asphalt

78 ASPHALT CONCRETE MIX DESIGN

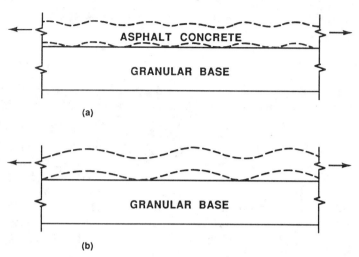

FIG. 1—*Comparison of original and polymer modified asphalt:* (a) *moderate viscosity and mix stiffness, and* (b) *high viscosity and mix stiffness.*

binder will essentially eliminate any creep relaxation of stresses and both wave length and amplitude will increase, as shown in Fig. 1b. However, prior application of heavy truck traffic during cooling will reduce the maximum amplitude attained as the pavement temperatures approach their maximum stiffness ($T \rightarrow T_g$). Obviously, the bond and friction between the asphalt concrete and granular base will have an influence on the magnitude or severity of contraction stresses and rippling.

The evidence that thermal rippling occurs in flexible pavements has been documented by tests conducted at the Florida Department of Transportation (FDOT), Bureau of Materials and Research, test pit facility [3]. Rapid depression of pavement temperatures from 20 to 0°C in the 2.4-m-wide by 3.7-m-long test pit pavement produced a rippling effect. Dual wheel load tests at pavement lift-off positions gave measured deflections and strains that were over twice those measured where the pavement was in contact with the base course. Why does lift-off occur when it is assumed that bonding to the base course is provided by a good prime coat? Usually, the prime does not penetrate sufficiently to produce adequate resistance of the granular base to the lift-off (rippling) forces. Similarly, inadequate tack and bonding of asphalt concrete layers can yield a similar effect. Once lift-off has developed, the frictional restraint of the pavement to contraction forces (strains) is minimal or perhaps non-existent after numerous repetitive thermal cooling cycles. Consequently, initial transverse cracking may occur at relatively large longitudinal distances (for example, 30 m). After initial cracking and release of tensile contraction stresses, the remaining uncracked pavement segment will contract and increase the amplitude of the ripples. This increases load-induced deflections, strains, and stresses that may result in a substantial increase in the amount of transverse cracking (closer crack spacing).

Surface ride measurements (Mays Meter, etc.) in Florida, Georgia, and Canada indicate that pavement roughness increases during the winter. In 1982, J. Hode-Keyser commented that analysis of Mays Meter data showed a smoother ride after pavement cracking. In Connecticut, J. E. Stephens has observed the rippling effect during fast warm up of the cold pavement. Further verification of flexible pavement rippling was documented by F. Hugo [4] using surface profile measurements before and after cooling of the pavement.

Effect of Asphalt and Mix Properties on Pavement Cracking

The influence of asphalt properties on low-temperature pavement cracking has been investigated primarily by researchers in northern climates. Fromm and Phang [5] developed creep test procedures and criteria to identify the critical temperature of asphalt mixtures. The use of Penetration-Visity number (PVN) to select paving-grade asphalts for improved low-temperature performance has been suggested by McLeod [6]. Ruth, Bloy, and Avital [7] established correlations between asphalt viscosity and mix parameters (resilient moduli, creep, fracture, etc.) that were used in the development of a computer program (CRACK) to evaluate the combined effect of thermal and load-induced stresses. Many other approaches have been proposed but, in general, most do not directly use asphalt properties measured at temperatures corresponding to in-service conditions.

An excellent illustration of asphalt hardness effects on transverse cracking was presented by Keyser and Ruth [8]. Correlation between penetration of recovered asphalt and the number of transverse cracks for two levels of traffic indicated a rapid increase in rate of cracking when the asphalt binder age hardened to 40 penetration (pen.) and lower.

Hugo and Kennedy [9] adopted the CRACK computer program for analysis of pavements in South Africa to demonstrate the effect of asphalt binder hardening on the relative potential for pavement cracking at low temperatures. Their analyses indicated critical temperatures in the range of 7 to $-10°C$. Apparently, their concern over excessive age hardening was warranted since Benkelman Beam test data obtained by the FDOT indicated that pavement deflections generally decrease with time until pavement cracking occurred, followed by increased deflections. This cracking was associated with asphalt binder viscosities at 25°C in excess of 1.0 MPa·s [10].

The degree of age hardening and the effect of asphalt source, plant mix temperatures, and age of pavement were evaluated by Page et al. [11] using data collected from a temperature reduction study. The results verify the critical nature of the interactions between asphalt type, plant mix temperature, and air void content of dense-graded asphalt concrete pavements.

Roque et al. [12] extended the prior work of Ruth [7] to evaluate pavement mixture properties on the basis of constant power viscosity, η_j [13]. Resilient moduli were predicted using previously established relationships with viscosity and actual measured values. These predictions were used in the elastic layer analysis of a test pit pavement. The results show good correspondence between predicted and measured deflections and strains. Subsequent test results at different low temperatures verified the validity of this approach [3]. Furthermore, the results of this investigation identified that pavement stresses and strains were primarily dependent upon deflection basin characteristics and asphalt mix properties that were controlled by asphalt viscosity and air void content. Consequently, a field testing program was initiated to obtain deflection data for stress analysis of pavement test sections.

An extensive nondestructive testing (NDT) program conducted by K. Badu-Tweneboah (using Dynaflect, falling-weight deflectometer (FWD), constant power viscosity of recovered asphalt binders, and Bitumen Structures Analysis in Roads (BISAR)) resulted in the development of simplified layer modulus prediction equations for four-layer systems [14–16]. Stress analysis of five of these pavement sections demonstrate the importance of asphalt binder hardness, subgrade support, and base course quality. Table 3 summarizes the results obtained from BISAR analyses using a 107.5 kN (24 kip) single-axle loading. Low stresses and deflections were obtained on SR-26B and SR-15C. Test road sections on SR-24 and US-441 gave low deflections but very high stresses due to the hardness of the asphalt binder. Both sections of SR-80 exhibited high deflections and high stresses. Base

80 ASPHALT CONCRETE MIX DESIGN

TABLE 3—*Summary of pavement stress analysis at low temperatures.*

Test Road	Temperature, °C	NDT Device	Maximum Deflection		Subgrade Compression, %	Subgrade Strain,[a] %	AC Stress[b]		Percent of AC Failure Stress[d]
			mm	mils			kPa	psi	
SR-26B	−5	Dynaflect	0.330	13.0	83.0	0.0166	614	89.0	22.3
		FWD	0.282	11.1	80.4	0.0151	625	90.7	22.7
SR-24	−5	Dynaflect	0.206	8.1	44.9	0.0088	1255	182.0	45.5
		FWD	0.297	11.7	35.5	0.0105	1917	278.0	69.5
US-441	−5	Dynaflect	0.254	10.0	58.2	0.0155	1434	208.0	52.0
		FWD	0.315	12.4	51.8	0.0176	1841	267.0	66.8
SR-15C	+5	Dynaflect	0.404	15.9	82.4	0.0161	375	54.4	13.6
		FWD	0.363	14.3	71.3	0.0173	610	88.5	22.1
SR-80-1	+5	Dynaflect	0.813	32.0	43.4	0.0191	1427	207.0	51.8
SR-80-2	+5	Dynaflect	0.978	38.5	24.7	0.0206	2372	344.0	86.0
SR-80[c]	+5	Dynaflect[c]	0.668	26.3	50.6	0.0175	513	74.4	18.6

[a] Vertical compressive strain on top of subgrade layer.
[b] Tensile stress at bottom of AC layer.
[c] Base course modulus increased to 586 mPa (85.0 ksi).
[d] Failure tensile stress of 2758 kPa (400 psi).

course moduli were 3.1 kPa (45 000 psi) and 2.21 kPa (32 000 psi) for Sections 1 and 2, respectively. This contributed to the differences in deflection and stresses. Limerock base moduli obtained from investigations of other pavements were typically 5.86 kPa (85 000 psi). This value was substituted into the BISAR analysis for SR-80-2 that reduced the maximum deflection 32% and stresses about 78%. The initial effect of base course quality is evident in this comparative analysis.

Prior indirect tensile strength testing of pavement cores generally produce slightly lower values than 50-blow Marshall compacted specimens that typically fail at a maximum low-temperature stress of 2.76 MPa (400 psi). The percent stress in the asphalt concrete pavement given in Table 1 were computed using the computed stress as a percent of the maximum tensile strength to indicate relative potential for cracking. Load-induced stress levels above 60% are generally considered critical since thermal stresses will increase actual stress levels. Extensive block cracking of the US-441 pavement and minor hairline cracks on SR-24 were observed. The poor base combined with inadequate surface drainage on SR-80-2 had produced pavement cracking prior to the time when Dynaflect tests were performed. Cracking was not visually apparent on SR-80-1 but six months later cracking was observed on this section. Although this demonstrates the validity of mechanistic procedures, the ultimate analysis would also incorporate the effect of thermal contraction, rippling, and reduction in foundation support due to moisture content changes (for example, high water table).

Selection of Asphalts and Polymer Modification

Current asphalt specifications do not provide (1) low-temperature parameters that directly relate to the crack resistance of flexible pavements or (2) suitable parameters for mechanistic pavement analysis of thermal and load effects. However, the specifications do require test data relating to asphalt properties at mixing, compaction, and high in-service temperatures. Unfortunately, most agencies place excessive emphasis on specifying specific viscosity grades of asphalt cements for hot mix asphalt concrete rather than evaluate or specify their properties through a suitable range in temperature (high to low). It is recognized that asphalts from certain crudes give excellent performance. As an example, tests on an air blown 85 to 100 pen. asphalt cement (AC) indicated low-temperature susceptibility and very low age hardening. Indirect tension tests on mixtures prepared with this AC and the result of the CRACK program showed improved low-temperature behavior. Mix properties at temperatures 10°C lower than tests on mixtures prepared with different asphalts were equivalent, indicating improved low-temperature performance.

Obviously, the availability of asphalts conforming to rigid specifications would be very limited and not feasible except in unusual situations. However, a properly formulated specification would allow the use of harder asphalts (for example, AC-50) provided their temperature susceptibility was sufficiently low to minimize low-temperature pavement cracking and age hardening. This means relatively low viscosities must exist at the low in-service pavement temperatures. Asphalts that have considerable volatile loss in the Thin Film Oven Test (TFOT) or Rolling Thin Film Oven Test (RTF) tests and at the hot-mix plant may be found suitable even though they may not conform to specifications. Perhaps these asphalts could be provided by the producer at a higher viscosity grading without adversely altering their low-temperature viscosity.

Another potential approach to high and low temperature problems is with polymer modification. The major problem with any admixture, additive, or modification process is the potential for variations in binder and mix properties with changes in asphalt supplier. This problem can be minimized if the asphalt suppliers provide polymer modified asphalts which are formulated to conform to a rational specification. However, current testing procedures and specifications are not adequate for assuring the quality of either asphalt cement or polymer modified asphalt.

A simple example of deficiencies encountered in conventional test methods normally required by current asphalt specifications is illustrated in Fig. 2. Assuming that the AC-30 control conforms to the specification, the polymer modified AC-30 (actually an AC-90) would not meet the 60°C absolute viscosity requirement even though it meets all other specification requirements. However, as demonstrated by the viscosity-temperature curves, the polymer modified asphalt should provide better high and low temperature pavement performance. This is based on the assumption that it can be mixed and placed without difficulty.

Now let us assume that standard penetration, ductility, and mixture tests are performed at 25°C. The test results will be the same for both control and modified unless the elastic effects of the polymer modified asphalt are substantially greater than the control. Interpretation of the test result could be erroneous if it is concluded that the polymer had no effect on asphalt properties. Herein lies the danger of using test measurements at one specified temperature. Therefore, it is necessary to have a temperature profile of the test parameter (for example, viscosity) for correct interpretation of the properties of both conventional and polymer modified asphalts.

Elastic characteristics of asphalts can be evaluated using ball penetration resilience tests, according to ASTM Methods of Testing Joint Sealants, Hot-Poured, for Concrete and Asphalt Pavements (D 3407-78), or Schweyer Rheometer shear modulus and stiffness measurements. The asphalts used in Fig. 2 were evaluated by resilience tests at 25°C. Control and polymer modified asphalts gave +6.7 and +17.0% recovery, respectively. Although these results indicate a substantial increase in the elastic properties of the modified asphalt,

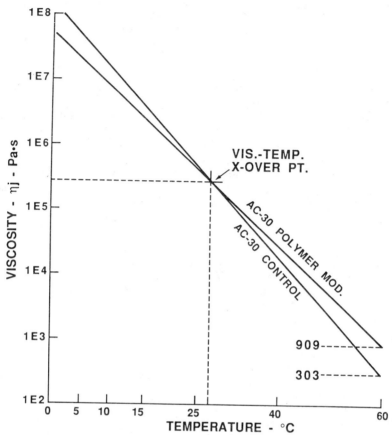

FIG. 2—*Differential temperature effects during rapid cooling at low temperatures.*

the empirical test values do not provide quantitative parameters for mechanistic analyses. However, shear moduli and time-stiffness comparison are quantitative. Test results on certain polymer-modified asphalts have shown equal or greater shear moduli even though the viscosity is lower than the control asphalt.

Schweyer Rheometer test results for certain polymer-modified asphalts indicate no delayed elastic response. Only the elastic and creep flow (two-component) response system exists that simplifies both testing and mechanistic analysis. The elimination of delayed elasticity provides better testing accuracy for viscosity and mixture creep tests. Also, it simplifies the analytical methods required for mechanistic analysis and design of asphalt pavements.

The indirect tensile strength of any given asphalt mixture is dependent upon asphalt viscosity. A higher viscosity (lower temperature) produces an increase in tensile strength up to the plateau value. Comparison of the tensile strengths of mixtures at equal asphalt viscosities for control and some polymer modified asphalts indicate the same tensile strength but greater failure strain and fracture energy for the polymer modified asphalts. Mixtures tested at equal binder viscosity temperatures provide a definitive characterization of asphalts or polymer modified asphalts or both. If no difference in results are obtained, then the modifier has done nothing more than harden or soften the binder without imparting any improvement in elastic behavior.

Suggestions for Asphalt Specifications

The test requirements for development of a model asphalt specification are given in Table 4. It is suggested that the original AC be characterized for solubility, flash point, smoke point, viscosity at 135 and 60°C, and penetration at 25°C. The TFOT, RFT, or similar procedures to obtain a residue representative of the pugmill or drum mix plant effects on asphalt hardening should be performed and the residue tested for viscosity (Item 2, Table 4), penetration, and specific gravity. These tests provide a characterization of the asphalt, similar to the Asphalt Residue (AR) grading, which should simulate as-constructed asphalt binder properties.

An age hardening procedure to simulate a 20-year-old asphalt should be developed and used to process AC for low-temperature performance predictions (Items 2 and 4, Table 4). Either TFOT or RFT may be suitable for this purpose. Another approach is to use laboratory-compacted specimens subjected to an age hardening process followed by recovery of the AC and testing according to that stipulated in Table 1.

Obviously, as proposed, it will require limits on the test parameters to establish a rational specification. Mechanistic analyses may provide alternates in material quality that means that a specification could be developed to incorporate two or more parameter levels or categories (for example, load, traffic volume, and temperature) for design and rehabilitation of flexible pavements.

Summary and Conclusions

An attempt has been made to provide a generalized overview of those factors that influence the hot and cold-temperature performance of asphalt concrete pavements. Emphasis

TABLE 4—*Test requirement recommendations for development of model asphalt specifications.*

1. Viscosity of original AC
   135°C   60°C poises or Pa·s
2. Viscosity after TFOT or RFT[a]
   60°C poises or Pa·s   5°C $\eta_{1.0}$, η_j, Pa·s
   25°C $\eta_{1.0}$, η_j, Pa·s   −5°C[c]
   15°C $\eta_{1.0}$, η_j, Pa·s
3. Viscosity after age hardening[b]
   25°C $\eta_{1.0}$, η_j, Pa·s   5°C $\eta_{1.0}$, η_j, Pa·s
   15°C $\eta_{1.0}$, η_j, Pa·s   −5°C[c] $\eta_{1.0}$, η_j, Pa·s
4. Shear moduli (Schweyer Rheometer, G tube)[b]
   5°C G, t_c, and S_j at 1.0s   −5°C[c]
5. Penetration at 25°C, (100 gm, 5 s) for reference use original AC, after TFOT or RFT, and after age hardening[b]
6. Solubility: original AC
7. Flash point test: original AC
8. Smoke point: original AC
9. Specific gravity of TFOT or RFT residue

[a] At a temperature representing plant mix temperatures that should simulate early (0 to 60 days) hardening of the AC.
[b] Asphalt age hardening procedure to simulate 20-year in-service hardened asphalt.
[c] Optional test temperature; lower temperature may be substituted, if desired.

has been placed on the development of rational specifications and mix design procedures to provide material parameters for incorporation into mechanistic pavement analysis and design systems.

Mix design procedures should primarily address the selection of suitable aggregates for blending to meet the desired gradation that will yield a design mixture having adequate resistance against hot weather consolidation rutting and plastic deformation. Data for a tentative gyratory mix design procedure was presented to illustrate the desired compaction and traffic simulation process for evaluation of the adequacy of the aggregate blend and for selection of a design asphalt content.

Asphalt cement properties primarily control the low-temperature behavior and performance of flexible pavements. Comments relating to the effects of asphalt viscosity, age hardening, thermal contraction and rippling, and wheel loading have been presented with emphasis on the use of asphalt viscosity for prediction of pavement stresses using NDT deflections and elastic layer (BISAR) analyses. This interactive mechanistic approach can eventually provide a means of achieving material characterization parameters for design of new pavements or the rehabilitation of existing pavements. Asphalt specifications should be developed according to those recommended for conventional acceptance or rejection purposes and for direct input into mechanistic procedures.

Specific conclusions derived from this assessment of current state-of-the-art information include the following.

1. Conceptually existing mix design procedures do not pragmatically provide information necessary to evaluate as-constructed and in-service rutting resistance.
2. Conceptually current asphalt test methods and specifications do not provide parameters to:
    (a) encompass the range of service temperatures,
    (b) evaluate the effects of age hardening, and
    (c) determine elastic behavior.
3. Current asphalt and mixture testing procedures need to be upgraded to provide parameters that relate directly to the behavior of asphalt concrete that can be easily used in mechanistic pavement analysis and design procedures.
4. Testing and data analysis procedures for evaluation of polymer modified asphalt are generally not adequate to clearly define the influence of the modifier. Current specification must be altered to accommodate polymer and similar additives for modification of asphalt cements.

References

[1] Ruth, B. E. and Schaub, J. H., "Gyratory Testing Machine Simulation of Field Compaction of Asphaltic Concrete," *Proceedings,* The Association of Asphalt Paving Technologists, Vol. 35, 1966, pp. 451–480.
[2] Ruth, B. E. and Schaub, J. H., "A Design Procedure for Asphaltic Concrete Mixtures," *Proceedings,* The Association of Asphalt Paving Technologists, Vol. 37, 1988, pp. 200–225.
[3] Ruth, B. E., Rogue, R., and Hardee, K. H., *Low-Temperature Pavement Response: Full Scale Pavement Testing,* Final Report Project 245-D34, Department of Civil Engineering, University of Florida, 1986, pp. 1–442.
[4] Hugo, F., Servas, V. P., and Snyman, D. R. F., "HVS-Aided Validation of Pavement Behavior at Low Temperature," *Proceedings,* The Association of Asphalt Paving Technologists, Vol. 56, 1987, p. 194.
[5] Fromm, H. J. and Phang, W. A., "A Study of Transverse Cracking of Bituminous Pavements," *Proceedings,* The Association of Asphalt Paving Technologists, Vol. 41, 1972, pp. 383–418.
[6] McLeod, N. W., Discussion, *Proceedings,* The Association of Asphalt Paving Technologists, Vol. 51, 1982, pp. 93–101.

[7] Ruth, B. E., Bloy, L. A. K., and Avital, A. A., "Prediction of Pavement Cracking at Low Temperatures," *Proceedings,* The Association of Asphalt Paving Technologists, Vol. 51, 1982, pp. 53–90.
[8] Keyser, J. H. and Ruth, B. E., "Analysis of Asphalt Concrete Test Sections in the Province of Quebec, Canada," *Transportation Research Record 968,* Transportation Research Board, 1984, pp. 54–65.
[9] Hugo, F. and Kennedy, T. W., "Surface Cracking of Asphalt Mixtures in South Africa," *Proceedings,* The Association of Asphalt Paving Technologists, Vol 54, 1985, pp. 454–496.
[10] Potts, C. F., Schweyer, H. E., and Smith, L. L., "An Analysis of Certain Variables Related to Field Performance of Asphaltic Pavements," *Proceedings,* The Association of Asphalt Paving Technologists, Vol. 42, 1973, pp. 564–585.
[11] Page, G. C., Murphy, K. H., Ruth, B. E., and Roque, R., "Asphalt Binder Hardening-Causes and Effects," *Proceedings,* The Association of Asphalt Paving Technologists, Vol. 54, 1985, pp. 140–160.
[12] Roque, R., Tia, M., and Ruth, B. E., "Asphalt Rheology to Define the Properties of Asphalt Concrete Mixtures and the Performance of Pavements," *ASPHALT RHEOLOGY—Relationship to Mixture ASTM STP 941,* American Society for Testing and Materials, Philadelphia 1985, pp. 3–27.
[13] Tia, M. and Ruth, B. E., in *Asphalt Rheology—Relationship to Mixture ASTM STP 941,* American Society for Testing and Materials, Philadelphia 1985, pp. 118–145.
[14] Ruth, B. E. and Badu-Tweneboah, K., "Non-Destructive Testing for the Structural Characterization of In-Place Pavement Materials," Final Report Project 245-D29, Department of Civil Engineering, University of Florida, 1986, pp. 1–114.
[15] Ruth, B. E., Tia, M., and Badu-Tweneboah, K., "Structural Characterization of In-Place Materials by Falling Weight Deflectometer," Final Report Project 245-D51, Department of Civil Engineering, University of Florida, 1986, pp. 1–199.
[16] Ruth, B. E., Puyana, E., and Badu-Tweneboah, K., "Pavement Layer Moduli Evaluation Using Dynaflect," *Proceedings,* 2nd International Conference on Bearing Capacity of Roads and Airfields, WDM Limited, Bristol, UK, 1986, pp. 299–308.

Gilbert Y. Baladi[1] and Ronald S. Harichandran[1]

Asphalt Mix Design and the Indirect Test: A New Horizon

REFERENCE: Baladi, G. Y. and Harichandran, R. S., "**Asphalt Mix Design and the Indirect Test: A New Horizon**," *Asphalt Concrete Mix Design: Development of More Rational Approaches, ASTM STP 1041,* W. Gartner, Jr., Ed., American Society for Testing and Materials, Philadelphia, 1989, pp. 86–105.

ABSTRACT: The features of a new indirect tension test apparatus is introduced. Analytical models to reduce the test data and to calculate the structural properties of asphalt mixes are presented and discussed. A summary of the findings along with the resulting statistical equations are also presented. It is shown that the structural properties of asphalt mixes obtained from the indirect tension test using the new apparatus are consistent and the test data are reproducible.

KEY WORDS: asphalt mixes, structural properties, indirect tension tests, constant load tests, cyclic load tests, mix design, asphalt concrete, asphalt specifications

During the past few years, the design of flexible pavement has rapidly evolved from empirical and semi-empirical methods to pavement design systems based on elastic or viscoelastic theories [1–3] or both. Today, many agencies use such systems in one form or another for new pavement and overlay designs. The use of such systems however, requires a thorough knowledge of the fundamental mechanical properties of the asphalt pavement materials that are a function of the asphalt mix variables [4–13]. Existing tests do not provide the properties required by these design procedures. Consequently, a variety of laboratory tests and testing equipment have been developed and employed [14]. Regardless of the complexity of the tests and testing equipment, it was found that different tests yielded different results and that the repeatability of the test results is highly questionable. Further, existing asphalt concrete mix design procedures are based upon empirical parameters that have no relationships to the structural design of asphalt pavements.

The Marshall method of asphalt mix design is a commonly used method for achieving mixtures with proper characteristics to be durable, stable yet pliable, and workable [15]. The method is based on two principles: (1) a density and voids analysis, and (2) a stability and flow test. Once Marshall tests are completed, the optimum asphalt content (the design asphalt content for the mixture) is determined based on a preselected criterion. Available criteria to select the best asphalt content are empirical and have no relationship to the structural properties of the mix. Therefore, after the optimum asphalt content is selected and the final mix is made, tests are conducted to determine the structural properties of that mix and consequently to design the pavement section. The problem in this procedure is that the final asphalt mix is accepted prior to the assessment of its structural properties.

---

[1] Professor and assistant professor, Civil Engineering, Department of Civil and Environmental Engineering, Michigan State University, East Lansing, MI 48824-1212.

From an engineering viewpoint, asphalt mix design should be based on the optimization of its structural properties to yield the best pavement performance under traffic loads and environmental conditions.

Recognizing the need to tailor the asphalt mix design procedure to the optimum structural properties and to be able to obtain these parameters from simple tests, this research project, sponsored by the Federal Highway Administration (FHWA), was undertaken. The objectives of the study included:

1. The selection of a simple test and test procedure that will allow the highway engineer to determine the fundamental engineering properties required for the structural design of asphalt pavements.
2. A study of the repeatability of the test results and the number of tests required to reliably obtain the mechanical properties (resilient modulus and Poisson's ratio, fatigue life and permanent deformation characteristics, creep, and viscoelastic properties) of the asphalt materials.

In this paper, a new indirect tension test apparatus is presented along with the analytical models to reduce the test data. Statistical models relating structural properties to the mix and test variables are also presented and briefly discussed.

Test Selection

To accomplish the objectives of this study, an extensive literature review of available tests and test procedures was undertaken. The findings have indicated that the triaxial and indirect tension tests are widely used for assessing the structural properties, and the flexural tests for estimating the fatigue life of the mix [16–28]. Consequently, in this study, several simple as well as sophisticated tests were employed. These included: triaxial tests (constant and repeated cyclic loads), cyclic flexural tests, Marshall tests, indirect tension tests (constant and variable cyclic loads), and creep tests. The test data indicated that:

1. The repeatability of test results is poor.
2. The material properties obtained from the different tests are substantially different.
3. The results from the indirect tension test were the most promising, although they were not consistent.

The last observation was made after examining the results of 24 tests that were conducted using existing apparatus (Schmidt's). The tests were conducted in triplicate using 45 and 226 kg (100 and 500 lb) cyclic loads. During the tests, the elastic (resilient), total, and plastic (permanent) deformations along the vertical and horizontal diameters of the test specimen were measured. It should be noted that the vertical deformation was measured using an extra linear variable differential transducer (LVDT) that was added to the Schmidt's apparatus. Nevertheless, the measured data are listed in Table 1. The resilient and total moduli were calculated using two methods as recommended by the ASTM Method for Indirect Tension Test for Resilient Modulus of Bituminous Mixtures (D 4123-82).

Method 1—In this method, the measured resilient and total horizontal deformations and an assumed value of Poisson's ratio of 0.35 were used. The results are listed in Table 2 as M_{R1} and E_1, respectively.

Method 2—In the second method, the measured resilient and total deformations along the vertical and horizontal diameters were used to calculate the resilient and total modulus and Poisson's ratio. The results are also listed in Table 2 as M_{R2} and E_2, respectively.

TABLE 1—*Indirect cyclic load data using the Schmidt apparatus.*

Sample Number	CL,[a] lb.	GMM[b]	AV,[c] %	Deformations[d] × 0.0001, in.[e]					
				Vertical			Horizontal		
				ELA	TOT	PLA	ELA	TOT	PLA
11115513	100	2.55	3.01	4.2	4.7	19.4	0.3	0.3	1.3
11115523	100	2.55	3.09	2.1	2.4	10.1	0.4	0.4	1.5
11115533	100	2.55	2.99	2.2	2.6	10.6	0.4	0.5	1.8
11115613	100	2.55	3.35	2.4	2.9	12.2	0.3	0.3	1.4
11115623	100	2.55	3.37	1.1	1.3	5.5	0.3	0.4	1.5
11115633	100	2.55	3.29	0.9	1.0	4.3	0.4	0.4	1.7
11115713	100	2.55	5.15	1.1	1.4	5.8	0.4	0.5	2.0
11115723	100	2.55	5.10	1.5	2.0	8.2	0.5	0.5	2.0
11115733	100	2.55	5.12	3.4	3.9	16.0	0.6	0.6	2.6
11115513	100	2.55	6.85	1.9	2.9	12.0	0.5	0.5	2.1
11115523	100	2.55	6.80	3.9	5.5	22.7	0.5	0.5	2.1
11115533	100	2.55	6.89	3.8	5.4	22.4	0.5	0.8	3.1
11115513	500	2.55	3.11	34.1	37.9	159.2	2.4	2.6	10.6
11115523	500	2.55	3.06	15.3	18.0	75.5	2.6	2.8	11.3
11115533	500	2.55	3.00	17.1	20.1	84.7	3.0	3.6	14.4
11115613	500	2.55	3.48	17.7	21.6	90.9	2.3	2.5	10.3
11115623	500	2.55	3.46	8.4	9.7	40.6	2.5	2.8	11.2
11115633	500	2.55	3.45	8.2	9.8	41.1	3.3	4.0	15.9
11115713	500	2.55	4.85	5.2	6.4	27.0	2.0	2.2	9.0
11115723	500	2.55	4.80	8.1	10.9	46.0	2.7	2.7	11.0
11115733	500	2.55	4.80	17.1	19.4	81.4	3.1	3.2	13.1
11115513	500	2.55	7.15	11.3	17.1	71.8	2.8	3.1	12.6
11115523	500	2.55	7.18	24.0	33.6	141.1	3.1	3.2	12.7
11115533	500	2.55	7.12	36.6	51.2	215.4	5.0	7.4	29.6

[a] CL = cyclic load, 1 lb = 0.45 kg.
[b] GMM = maximum theoretical specific gravity.
[c] AV = percent air voids.
[d] ELA and TOT = elastic and total deformation/cycle, and PLA = cumulative plastic (permanent) deformation.
[e] 1 in. = 2.54 cm.

Examination of the values of the resilient and total moduli listed in Table 2 indicates the following.

1. The values of the resilient modulus obtained using Method 1 are different by as much as a factor of 2 or better than those obtained using Method 2.
2. The values obtained for the resilient modulus from a triplicate vary by a factor of 1.9.
3. For all tests, the values of the total modulus (from Method 1 or 2) are smaller than the values of the resilient modulus.
4. The values of the resilient and total Poisson's ratios vary from −0.02 to 1.18.
5. In general, decreasing percent air voids yield an increase in the value of M_{R1}, M_{R2}, E_1, and E_2 (see Fig. 1).
6. Increasing cyclic load results in a decrease in all moduli values.

The first two observations imply that the test results are not consistent. The other four observations were expected although the trend is not consistent.

In addition, during the tests, it was noticed that the measured horizontal and vertical

TABLE 2—*Resilient and total moduli and Poisson's ratios from indirect tension tests using the Schmidt apparatus.*

Sample Number	AV,[a] %	CL,[b] lb	Moduli,[c] psi				Poisson's Ratio	
			M_{R1}	M_{R2}	E_1	E_2	Resilient	Total
11115513	3.01	100	835243	340939	759974	307302	−0.02	−0.02
11115523	3.09	100	702123	694723	650114	590575	0.34	0.29
11115533	2.99	100	651235	664186	546271	562474	0.36	0.37
11115613	3.35	100	775234	593354	713215	486998	0.21	0.15
11115623	3.37	100	737225	1250833	647538	1083703	0.78	0.77
11115633	3.29	100	705263	1651879	589027	1381301	1.18	1.18
11115713	5.15	100	568235	1256379	505651	1022009	1.10	0.98
11115723	5.10	100	512346	981096	500795	726262	0.92	0.63
11115733	5.12	100	396455	419767	382509	371507	0.39	0.33
11115513	6.85	100	531245	750263	469011	496142	0.61	0.39
11115523	6.80	100	496578	367850	481405	262695	0.19	0.07
11115533	6.89	100	467244	373096	320235	266497	0.23	0.25
11115513	3.11	500	515244	210318	468812	189568	−0.02	−0.02
11115523	3.06	500	475235	470226	440032	399733	0.34	0.29
11115533	3.00	500	412578	420783	346080	356345	0.36	0.37
11115613	3.48	500	528765	404710	486464	332168	0.21	0.15
11115623	3.46	500	505698	858006	444178	743364	0.78	0.77
11115633	3.45	500	375190	878776	313354	734833	1.18	1.18
11115713	4.85	500	620987	1373015	552593	1116887	1.10	0.98
11115723	4.80	500	462777	886176	452344	655996	0.92	0.63
11115733	4.80	500	395690	418957	381771	370790	0.39	0.33
11115513	7.15	500	450129	635705	397397	420386	0.61	0.39
11115523	7.18	500	404235	299445	391883	213845	0.19	0.07
11115533	7.12	500	245691	196185	168389	140132	0.23	0.25

[a] AV = percent air voids.
[b] CL = cyclic load, 1 lb = 0.45 kg.
[c] M_{R1} = calculated resilient modulus using Poisson's ratio of 0.35 and the horizontal elastic deformation; M_{R2} = calculated resilient modulus using horizontal and vertical elastic deformations; and E_1, E_2 = as M_{R1} and M_{R2} but total modulus.

deformations were dependent on the placement of the specimen on the lower curved platen (two sequential placements yielded different measurements) and that the upper head of the instrument experienced a slight rocking motion. Hence, the inconsistency in the indirect tension test results seemed to be related to the equipment rather than the test mode. Existing indirect test apparatus have one or more of the following problems.

1. A rocking motion of the loading head that distorts the accuracy of the vertical deformation of the test specimen.
2. Due to equipment configuration, the test specimen may roll over the lower curved platen resulting in erroneous horizontal deformations.
3. The position of the test specimen on the lower curved platen is arbitrary and differs from one specimen to another.
4. During a repeated load test, the horizontal axis of the test specimen may rotate relative to the vertical axis of the loading head resulting in a smaller measurement of the horizontal deformation on one side of the diameter relative to the opposite side.
5. The lack of the capability to measure specimen deformations in three directions indicates that some information available from the test cannot be recorded.

90 ASPHALT CONCRETE MIX DESIGN

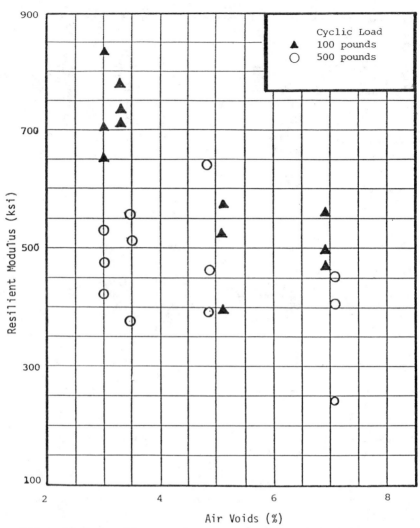

FIG. 1—*Resilient modulus versus the percent air voids for asphalt mix specimens tested using existing indirect tension test apparatus.*

Recognizing these problems, a new indirect tension test apparatus was designed. A brief summary of some of the features of that apparatus is presented next.

New Indirect Tension Test Apparatus

As just noted, to overcome the problems associated with existing apparatus, a new simple and inexpensive indirect tension test apparatus was designed (by the first author) and was later modified and fabricated by personnel at Michigan Department of Transportation (MDOT). The new apparatus has the following features.

1. The deformation of the indirect test specimen can be measured in one, two, or three directions using either one or two LVDTs in each direction.
2. The apparatus can be used under any existing loading frame (for example, Marshall, hydraulic system, unconfined, triaxial) and it has a guiding system that consists of four frictionless pistons.
3. The function of the frictionless guiding system is to prevent rotation or rocking or both of the upper curved loading strip of the apparatus.
4. The apparatus has four reference positions enabling it to be easily placed under the center of the loading mechanism of a standard loading frame.
5. The apparatus has a sample stopper for easy positioning of the test specimen on the lower loading strip of the apparatus and for perfect alignment of the horizontal diameter (axis) of the specimen with the axis of the horizontal LVDT(s).
6. The apparatus was made using parts that are available at any typical machine shop or supply store.

In the following sections, the design of this new apparatus is introduced. Observations concerning the new apparatus, the test results, and the analytical models (equations) to reduce the test data are also presented.

Engineering Drawings

The engineering drawings of the new indirect tension test apparatus are shown in Figs. 2 and 3. Figure 2 shows an elevation view of the assembled apparatus and specimen stopper, and one top view and several elevation views of the lower stationary plate. The materials list is also presented on this sheet. Figure 3 shows top and elevation views of parts of the apparatus.

One copy of the apparatus was manufactured by the machine shop personnel at MDOT. This copy was utilized in the laboratory testing of this study. During the course of the investigation, it was noted that *the friction between the loading piston and the upper stationary plate is minimal.* This is mainly due to the workmanship of the machine shop personnel. If the friction between these two parts of the apparatus is of concern, then *a load cell can be added between the lower curved load strip and the lower stationary plate.* The load recorded by the load cell will then be the true load applied to the test specimen.

It should be noted that two other copies of the apparatus were made. One was loaned to the University of Texas A & M for utilization on an National Cooperative Highway Research Program (NCHRP) sponsored project. The other was for use by the MDOT laboratory.

Analytical Models for the Indirect Tension Test

The analytical models used to extract the resilient and total moduli, Poisson's ratio, and the compressive and tensile strength from the measured deformations, are based on the assumptions that compacted asphalt mixes are homogeneous, isotropic, and linear elastic. Detail relating to the development of the model can be found in a final report to the FHWA [29] and in Ref 24. The equations to calculate the resilient and total moduli and Poisson's ratios, and the compressive and tensile strengths for 10.2 and 15.2-cm (4 and 6-in.) diameter specimens are given below in Eqs 1 through 5.

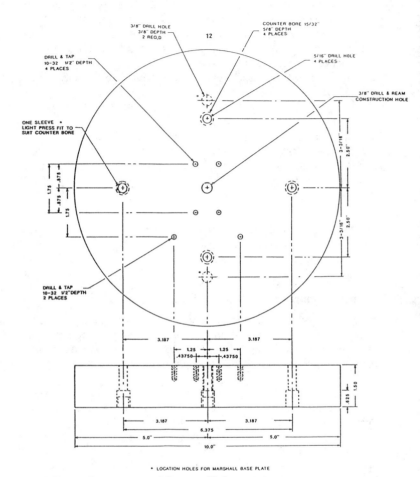

FIG. 2—*Elevation view of the assembled apparatus* (Part 1).

$$U = [A1 - (A2)(DR)]/[A3 + DR] \qquad (1)$$

$$M_R \text{ or } E = P[A1 - (A3)(U)]/[L(D_V)] \qquad (2)$$

$$M_R \text{ or } E = [(A4)(P)(U)]/[D_L] \qquad (3)$$

$$INCS = (A5)(P)/(L) \qquad (4)$$

$$INTS = (A6)(P)/(L) \qquad (5)$$

where

U = Poisson's ratio;
DR = deformation ratio = DV/DH;

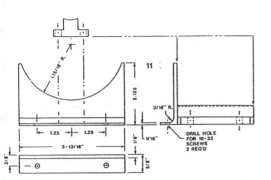

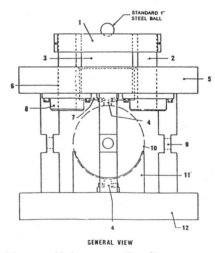

FIG. 2—*Elevation view of the assembled apparatus* (Part 2).

D_V = vertical resilient or total deformation of the specimen along the vertical diameter, in.;
D_H = horizontal resilient or total deformation of the specimen along the horizontal diameter, in.;
M_R = resilient modulus, psi;
E = total modulus, psi;
L = sample thickness, in.;
D_L = longitudinal deformation along the longitudinal axis (thickness) of the specimen, in.;
P = the magnitude of the applied load, lb;
INCS = maximum compressive stress at the center of the test specimen, psi;

94 ASPHALT CONCRETE MIX DESIGN

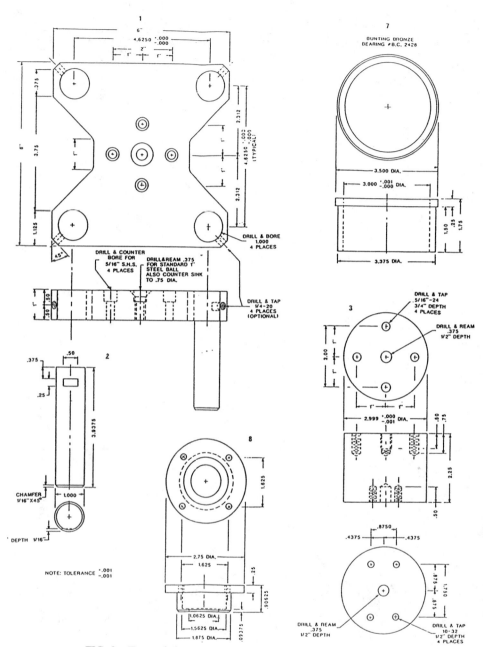

FIG. 3—*Top and elevation views of parts of the apparatus* (Part 1).

BALADI AND HARICHANDRAN ON INDIRECT TENSION TEST APPARATUS 95

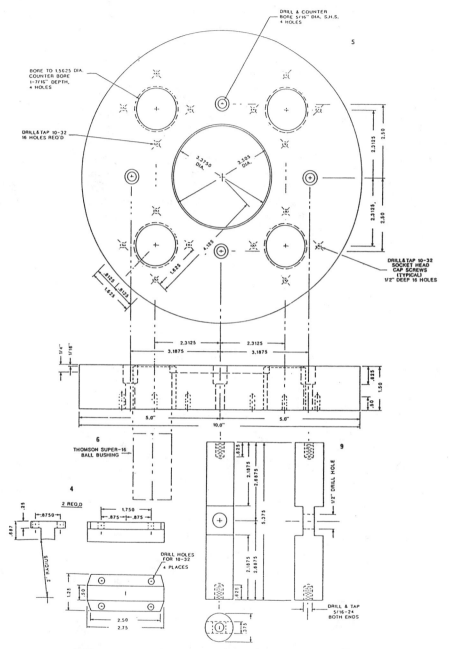

FIG. 3—*Top and elevation views of parts of the apparatus* (Part 2).

INTS = maximum tensile stress at the center of the test specimen, psi; and $A1$, $A2$, $A3$, $A4$, $A5$, and $A6$ are integration constants; the value of these constants for the 10.2 and 15.2-cm (4 and 6 in.) specimens are given in Table 3.

Theoretically, the values of the resilient modulus from Eqs 2 and 3 should be exactly the same for homogeneous, isotropic, and linear elastic material. Asphalt mixes are heterogeneous and anisotropic. Due to this and measurement errors, differences between the two calculated values should be expected. Equation 2 can be utilized if the deformation in the third dimension is not measured. If the sample deformations in all three directions are measured, then the two values of the calculated resilient modulus from Eqs 2 and 3 should be compatible (a maximum difference of 5% was noted in this study). Substantial difference between these two values may mean that the test results are not accurate. A better procedure, however, is to estimate the modulus and Poisson's ratio using all measurements such that the sum of the errors between the theoretically predicted and measured deformations is minimized. This is known as the least squares method that was performed for the 10.2-cm (4-in.) specimen and resulted in the following equations.

$$U = (0.225127 \times H^2 - 0.269895 \times V^2 - 0.0447676 \times A^2 + 3.570975 \times H \times V + 0.086136 \times A \times H + 1.145064 \times A \times V)/D \tag{6}$$

$$M_R \text{ or } E = (0.253680 \times H + 3.9702876 \times V - 0.0142874 \times A)/D \tag{7}$$

where

$D = 1.105791(H^2 + V^2 + A^2) - (H - 0.0627461 \times V + 0.319145 \times A)^2$;
$H = D_H \times L/P$;
$V = D_V \times L/P$;
$A = D_L/P$; and
M_R, E, D_H, D_V, D_L, L, and P are as before.

If the sample deformations are measured in all three directions, then Eqs 6 and 7 yield the best estimates of the resilient modulus and Poisson's ratio. If one of these three deformations, however, is not available, then Eqs 1 and 2 or 3 should be used.

TABLE 3—*Values of the constant of integration for 10.2 and 15.2-cm (4 and 6-in.) diameter specimens.*

	Specimen Diameter	
Constant	4 in.[a]	6 in.
A1	3.587910	4.085950
A2	0.269895	0.271760
A3	0.062745	0.041733
A4	0.319145	0.212453
A5	0.475386	0.105242
A6	0.156241	0.317695

[a] 1 in. = 2.54 cm.

Implementation

The frictionless boundary model (Eqs 6 and 7) was utilized in this study to calculate the resilient and total moduli and Poisson's ratios using measured deformations in three directions. For the resilient modulus and Poisson's ratio, the resilient deformation was used, while the total deformations were used for the total modulus and Poisson's ratio. The calculated values of the resilient and total moduli and Poisson's ratios were then correlated to the mix and test variables. The resulting equations are presented in the section on statistical models.

Laboratory Tests Using the New Apparatus

The following tests were conducted using the new indirect tension test apparatus.

1. Indirect tension tests (INTT) utilizing a standard Marshall loading frame and deformation rate. In this test mode, some of the test specimens were conditioned as standard Marshall specimens. Others were tested at 25 and 5°C (77 and 40°F).
2. Indirect constant peak cyclic load (INCCL) tests utilizing an MTS hydraulic system. In this test, the specimens were subjected to a constant sustained load followed by a constant peak cyclic load of 226 kg (500 lb). Some of the test specimens were subjected to a maximum of 500 000 cycles at a frequency of two cycles per second with a loading time of 0.1 s and a relaxation period of 0.4 s. Measurements of the elastic, total, and plastic (permanent) deformations were collected along the vertical and horizontal diameters, and along the thickness of the specimen. The data were then analyzed to obtain the resilient and total characteristics of the specimens and their fatigue lives.
3. Indirect variable peak cyclic load (INVCL) test utilizing an MTS hydraulic system. Basically, the test procedure is the same as that of the INCCL except that after the application of the sustained load, the specimen was subjected to 45, 91, and 226 kg (100, 200, and 500 lb) peak cyclic loads with each load being applied for only 1000 cycles.

Results from the INTT and Eqs 4 and 5 were used to calculate the compressive and tensile strengths of the mixes. The data from the INCCL tests and Eqs 6 and 7 were utilized to obtain the resilient and total moduli and Poisson's ratios and the fatigue lives of the test specimens. Results from the INVCL were analyzed to infer the effects of the magnitude of the cyclic loads upon the values of the moduli and Poisson's ratios.

Test Specimens

A total of 412 indirect test specimens were fabricated and tested using the new indirect test apparatus. The specimens were made using the following materials.

(a) Three different types of aggregate (crushed and angular limestone, relatively rounded natural aggregate, and a mix of 50% by weight per sieve of the crushed limestone and natural aggregates).
(b) Two different aggregate gradations.
(c) Three viscosity graded asphalts (AC-2.5, AC-5, and AC-10).

98 ASPHALT CONCRETE MIX DESIGN

For each material combination, a constant percent asphalt content was utilized (the percent asphalt content at 3% air voids as determined from the standard Marshall mix design). The test specimens were made at three different percent air voids (3, 5, and 7%) by varying the compaction efforts. For each material combination and percent air void, one triaxial type sample, 10.2-cm diameter by 21.6-cm high (4-in. and 8.5-in.), was made. Later, the triaxial sample was cut into three 6.4 cm high (2.5 in.) indirect test specimens. The three indirect specimens (triplicate) were then tested under the same conditions (test temperature and test type) using the new indirect test apparatus.

Test Results

The test results (using the new indirect tension test apparatus) are tabulated in Ref 29. Examination of the results indicated the following.

1. The values of the resilient modulus obtained from the INCCL tests agree very well with those obtained from the INVCL tests.
2. Poisson's ratio was found to be a function of the mix variables. For all of the test specimens, the value of Poisson's ratio varies from 0.2 to 0.42; and for any triplicate, the values of Poisson's ratio are almost the same.
3. The fatigue lives of all test specimens are more meaningful, more reasonable, and more consistent (for a triplicate) than those obtained from flexural cyclic load tests.
4. The cumulative permanent (plastic) deformations (from cyclic load tests) of the test specimens can be analyzed in both compression and tension modes and they are more consistent than those obtained from cyclic load triaxial or flexural beam tests.
5. For any combination of variables, the test results can be reproduced with a high degree of accuracy.

For all tests, the resilient and total moduli and Poisson's ratio and the indirect compressive and tensile strengths were calculated using the measured data and the analytical models presented in the section on analytical models.

Statistical Models

As just noted, the values of the indirect compressive and tensile strengths and the resilient and total moduli and Poisson's ratios calculated using Eqs 4 through 7 and the measured data were statistically correlated to the mix and test variables. A detailed discussion of the statistical analysis may be found in Refs 29 and 30. These statistical analyses resulted in Eqs 8 through 13.

$$\ln(M_R) = 16.092 - 0.03658(TT) - 0.1401(AV) - 0.0003409(CL) + 0.04353(ANG) + 0.0008793(KV)$$
$$R^2 = 0.997;\ \text{and SE} = 0.033 \tag{8}$$

$$\ln(E) = 16.385 - 0.04529(TT) - 0.1549(AV) - 0.0003339(CL) + 0.4258(ANG) + 0.0008364(KV)$$
$$R^2 = 0.998;\ \text{and SE} = 0.034 \tag{9}$$

$$\ln(PT) = -0.43228 - 0.01940(TT) - 0.06329(AV) + 0.001332(KV) + 0.0001236(CL)$$
$$R^2 = 0.913;\ \text{and SE} = 0.108 \tag{10}$$

$$\ln(PR) = -1.370 - 0.04243(AV) + 0.000885(KV) + \\ 0.004662[1n(N)] + 0.0004489(TT) \quad (11) \\ R^2 = 0.720; \text{ and SE} = 0.045$$

$$\ln(INCS) = 9.1350 - 0.03369 \times TT - 0.2604 \times AV + \\ 0.05223 \times ANG + 0.007399 \times KV \quad (12) \\ R^2 = 0.99 \text{ and SE} = 0.08$$

$$\ln(INTS) = 8.01589 - 0.03363 \times TT - 0.2605 \times AV + \\ 0.0509 \times ANG + 0.0007676 \times KV \quad (13) \\ R^2 = 0.99 \text{ and SE} = 0.08$$

where

- ln = natural logarithm;
- M_R = resilient modulus, psi;
- E = total modulus, psi;
- PT = total Poisson's ratio;
- PR = resilient Poisson's ratio;
- INCS = indirect compressive strength, psi;
- INTS = indirect tensile strength, psi;
- TT = test temperature, °F;
- AV = percent air voids, (AV = 1, 2, 3, etc.);
- CL = applied cyclic load, lb;
- L = actual sample thickness, in.;
- ANG = aggregate angularity (angularity was assigned a scale from 1 to 4; a value of 1 represents perfectly spherical and smooth particles, while 4 represents highly angular particles); for this study the angularity of the crushed limestone is 4, the rounded natural aggregate is 2, and the angularity of the 50% mix by weight was given a value of 3;
- KV = kinematic viscosity of the asphalt 275°F (135°C), cSt, (AASHTO T201);
- CD1 = cumulative vertical plastic deformation, in.;
- GRAD = gradation factor (GRAD = 1 for Gradation A, and 0.98 for Gradation B); since only two types of gradation were used in this project, a meaningful relationship between the GRAD and the resilient modulus cannot be obtained;
- N = the number of load repetitions;
- R^2 = coefficient of correlation; and
- SE = standard error.

The sensitivity of the values of the resilient modulus predicted by Eq 8 to variations in the values of the independent variables (TT from 4.4 to 25°C (40 to 77°F); AV from 3 to 7%; CL from 45 to 226 kg (100 to 500 lb), ANG from 2 to 4, and KV from 159 to 270 cSt) was studied. The following points were noted.

(a) The values of the resilient modulus increase by a factor of 3.76 as the temperature decreases from 4.4 to 25°C (77 to 40°F).

(b) A decrease in the percent air voids from 7 to 3% results in an increase in the resilient modulus by a factor of 1.76.

(c) An increase in the applied load from 45 to 226 kg (100 to 500 lb) results in a reduction in the modulus by a factor of 1.15.

(d) The values of the resilient modulus increase by a factor of 1.09 as the ANG increases from 2 (rounded aggregate) to 4 (crushed aggregate).

(e) Increasing KV from 159 to 270 cSt leads to increasing M_R by a factor of 1.1.

These observations imply that the asphalt mix has a nonlinear elastic behavior (higher load yields lower modulus values). The significance of this is that existing standard test procedures (for example, ASTM D 4123-82) specifies that the resilient modulus tests be conducted using low magnitude of the cyclic load. This will result in higher estimates of the resilient modulus than at higher loads. A correct evaluation of the values of the total or resilient modulus should include the sensitivity of these values to the range of the load anticipated in the field (that is, obtain the relationship between M_R or E or CL). Further, it is stated in ASTM D 4123-82 that "If Poisson's ratio is assumed, the vertical deformations are not required. A value of 0.35 for Poisson's ratio has been found to be reasonable for asphalt mixtures at 77°F." The values of the resilient modulus in this study were also calculated (as specified by ASTM D 4123-82) using the measured horizontal deformation and an assumed value of Poisson's ratio of 0.35. It was found that the assumption of Poisson's ratio of 0.35 consistently overestimated the modulus values. An assumption of Poisson's ratio of 0.27 yields better estimates of the values of M_R for most data points. It should be noted that a change in the assumed value of Poisson's ratio of 0.01 results in a change in the value of M_R by about 2%. Since Poisson's ratio of asphalt mixes is a function of the mix and test variables (see Eqs 10 and 11), it is strongly recommended that Poisson's ratio be calculated using measured vertical and horizontal deformations.

Nevertheless, Fig. 4 shows a plot of the measured (Eq 7) and calculated (Eq 8) values of resilient modulus. The straight line in the figure represents equality between measured and calculated values. It should be noted that the maximum difference between the arithmetic (not logarithmic) values of the measured and calculated (using Eq 8) values is only 8%. A plot of the calculated and measured values of the total modulus showed a similar trend to that of Fig. 4.

Similarly, Figs. 5 and 6 show, respectively, plots of the measured (Eq 6) and calculated (Eq 11) values of the resilient Poisson's ratio and the measured (Eq 4) and calculated (Eq 12) values of the indirect compressive strength. The straight lines in the figures represent equality between measured and calculated values. Also shown in Fig. 5 is the value of Poisson's ratio of 0.35 as recommended by ASTM D 4123-82. It can be seen that the recommended value is much higher than the measured ones. Plots for the values of the total Poisson's ratio and the indirect tensile strength showed similar trends to those of Figs. 5 and 6, respectively.

It can be seen from Figs. 4 through 6 that the statistical equations are accurate. However, it is not recommended herein that these equations be used to estimate the values of the resilient modulus. The reason is that the equations are based upon a limited data base. Therefore, verification of the equations using actual laboratory tests is highly recommended.

Conclusions

Based on the laboratory test results and the analytical and statistical analyses, the following conclusions can be stated.

1. For all specimens tested using the new indirect tension test apparatus, the test results are consistent and very reasonable.

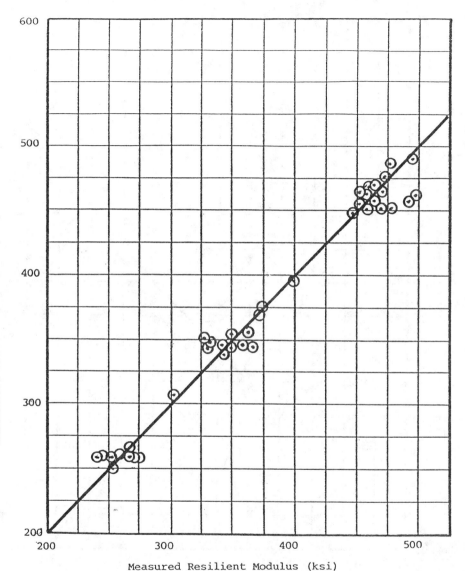

FIG. 4—*Measured and estimated values of the resilient modulus, the INCCL tests.*

2. For any triplicate tested using the new apparatus, the maximum difference between the results of the three tests is only 7%.
3. The resilient characteristics of asphalt mixes can be expressed in terms of the mix and test variables.
4. The test temperature and the percent air void in the mix have the greatest influence on the resilient characteristics of the mix.
5. The resilient modulus decreases as the number of load repetitions increases (softening effects).

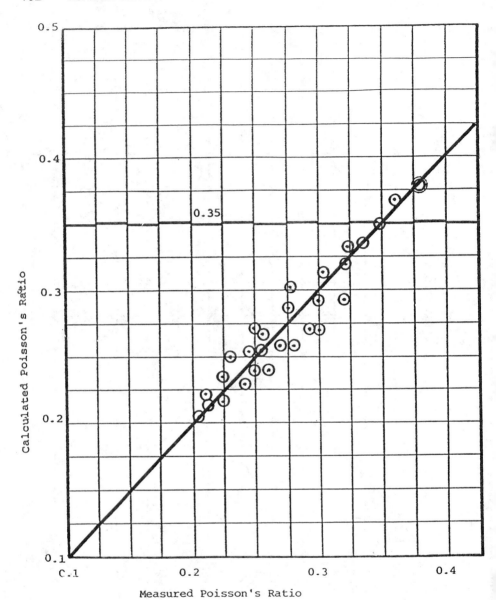

FIG. 5—*Measured and estimated values of the resilient Poisson's ratio, the INCCL tests.*

6. Poisson's ratio of the mixes is dependent on the air voids in the mix and the kinematic viscosity of the binder.
7. Increasing aggregate angularity causes an increase in the resilient modulus.
8. Higher asphalt binder viscosity produces stiffer mix and higher resilient characteristics.
9. For any test specimen, the test results are consistent and very reasonable.

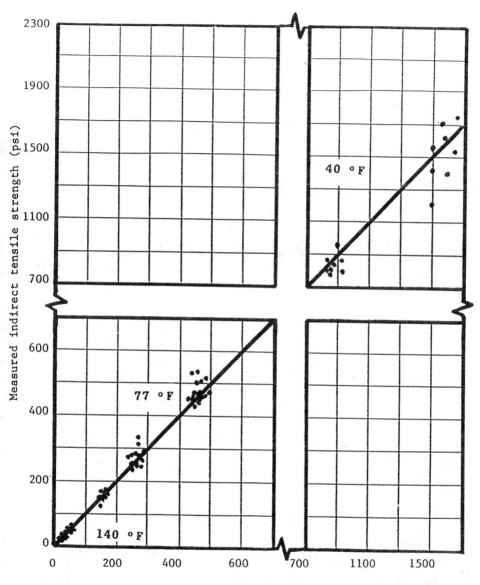

FIG. 6—*Measured and estimated values of the indirect compressive strength at 25 and 4.4°C (77 and 40°F) for the INTT tests.*

10. The maximum difference in the values of the resilient or total moduli obtained from a triplicate is only 7%.
11. The values of Poisson's ratio for all 414 specimens vary from 0.23 to 0.32. Further, for any triplicate, Poisson's ratio is almost constant.
12. For any combination of variables, the test results can be accurately reproduced.

13. The ASTM D 4123-82 test procedure is inadequate and may lead to misleading results.
14. The indirect tension tests can be used to obtain an asphalt mix design based on the structural properties of the mix.

Summary

Structural properties of asphalt mixes have direct bearing on the pavement performance. A correct estimate of these properties is essential for the structural design of pavements. Existing indirect tension test apparatus have several problems that cause an inconsistency in the test results. A new indirect tension test apparatus was designed, fabricated, and utilized in a study sponsored by the FHWA. Results obtained using the new apparatus were found to be consistent and reproducible.

Acknowledgment

The authors would like to express their deep gratitude to the FHWA for its financial support and to the personnel of the Division of Testing and Technology of MDOT for their understanding and valuable contributions.

References

[1] "AASHTO Interim Guide for Design of Pavement Structures," American Association for State Highway and Transportation Officials, 1972, Chapter III (revised), 1981.
[2] Baladi, G. Y., "Characterization of Flexible Pavement: A Case Study," *Properties of Flexible Pavement Materials, ASTM STP 807*, American Society for Testing and Material, J. J. Emery, Ed., Philadelphia, 1983, pp. 164–171.
[3] Kenis, W. J. in *Fatigue and Dynamic Testing of Bituminous Mixtures, ASTM STP 561*, American Society for Testing and Materials, Philadelphia, 1973, pp. 132–152.
[4] Miller, J. S., Uzan, J., and Witczak, M. W., "Modification of the Asphalt Institute Bituminous Mix Modulus Predictive Equation," Transportation Research Record 911, 1983, pp. 27–36.
[5] Nijboer, L. W., "Mechanical Properties of Asphalt Materials and Structural Design of Asphalt Roads," Highway Research Board, Vol. 33, 1954, pp. 185–200.
[6] Shook, J. F. and Kallas, B. F., "Factors Influencing Dynamic Modulus of Asphalt Concrete," The Association of Asphalt Paving Technologists, Vol. 38, 1969, pp. 140–178.
[7] Yoder, E. J. and Witczak, M. W., *Principles of Pavement Design*, 2nd ed., Wiley, New York, 1975.
[8] "Asphalt Overlays for Highway and Street Rehabilitation," The Asphalt Institute, Manual Series No. 17 (MS-17), June 1983.
[9] Brown, S. F. and Cooper, K. E., "A Fundamental Study of the Stress-Strain Characteristics of a Bituminous Material," The Association of Asphalt Paving Technologists, Vol. 49, 1980, pp. 476–499.
[10] Finn, F. N., "Factors Involved in the Design of Asphaltic Pavement Surfaces," National Cooperative Highway Research Program, Report No. 39, 1967, pp. 1–112.
[11] Kallas, B. F. and Shook, J. F., "Factors Influencing Dynamic Modulus of Asphalt Concrete," The Association of Asphalt Paving Technologists, Vol. 38, 1969, p. 140.
[12] Von Quintus, H. L., Rauhut, J. B., and Kennedy, T. W., "Comparisons of Asphalt Concrete Stiffness as Measured by Various Testing Techniques," The Association of Asphalt Paving Technologists, Vol. 51, 1982, pp. 35–52.
[13] Witczak, M. W. and Root, R. E. in *Fatigue and Dynamic Testing of Bituminous Mixtures, ASTM STP 561*, American Society for Testing and Materials, Philadelphia, 1974, pp. 67–94.
[14] "AASHTO Test and Material Specifications," Parts I and II, 13th ed., American Association for State Highway and Transportation Officials, 1982.
[15] "Mix Design Methods for Asphalt Concrete and other Hot-Mix Types," The Asphalt Institute, Manual Series No. 2 (MS-2), College Park, MD, 1979.

[16] Farzin, M. H., Krizek, R. J., and Corotis, R. B., "Evaluation of Modulus and Poisson's Ratio From Triaxial Tests," Transportation Research Record 537, 1975, pp. 69–80.
[17] Goetz, W. H., "Comparison of Triaxial and Marshall Test Results," The Association of Asphalt Paving Technologists, Vol. 20, 1951, pp. 200–245.
[18] Gonzalez, G., Kennedy, T. W., and Anagnos, J. N., "Evaluation of the Resilient Elastic Characteristics of Asphalt Mixtures Using the Indirect Tensile Test," Report No. CFHR 3-9-72-183-6, Transportation Planning Division, Texas State Department of Highways and Public Transportation, Austin, TX, Nov. 1975.
[19] Hadley, W. O. and Vahida, H., "A Fundamental Comparison of the Flexural and Indirect Tensile Tests," presented at the Transportation Research Board meetings, Jan. 1983.
[20] Hills J. F. and Heukelom, W. "The Modulus and Poisson's Ratio of Asphalt Mixes," *Journal,* Institute of Petroleum, Vol. 55, Jan. 1969, pp. 27–35.
[21] Kennedy, T. W., "Characterization of Asphalt Pavement Materials Using the Indirect Tensile Test," The Association of Asphalt Paving Technologists, Vol. 46, 1977, pp. 132–150.
[22] Monismith, C. L., "Flexibility Characteristics of Asphalt Paving Mixtures,'" The Association of Asphalt Paving Technologists, Vol. 27, 1958, pp. 74–106.
[23] Pell, P. S. and Brown, S. F., "The Characteristics of Materials for the Design of Flexible Pavement Structures," The 3rd International Conference on the Structural Design of Asphalt Pavements, University of Michigan, Vol. I, 1972, pp. 326–342.
[24] Timoshenko, S. P. and Goodier, J. N., *Theory of Elasticity,* McGraw Hill Book Company, New York, 1970.
[25] Young, M. A. and Baladi, G. Y., "Repeated Load Triaxial Testing, State of the Art," Michigan State University, Division of Engineering Research, 1977.
[26] Baladi, G. Y., "Linear Viscosity," U.S. Army Corps of Engineers Waterways Experiment Station, Oct. 1985, pp. 1–6.
[27] Baladi, G. Y., "Numerical Implementation of a Transverse-Isotropic, Inelastic, Work-Hardening Constitutive Model," Soil Dynamics Division, Soils and Pavement Laboratory, U.S. Army Corps of Engineers Waterways Experiment Station, Vicksburg, MS, pp. 1–12.
[28] Baladi, G. Y. and De Foe, J. H., "The Indirect Tensile Test: A New Horizon," an interim report submitted to the Federal Highway Administration, March 1987.
[29] Baladi, G. Y., "Integrated Material and Structural Design Method for Flexible Pavements," Final Report No. FHWA/RD-88/109, 110 and 118, Sept. 1987.
[30] Norusis, M., "SPSS/PC+ for the IBM PC/XT/AT," SPSS Inc., Chicago, 1986.

William O. Yandell[1] *and Robert B. Smith*[2]

Toward Maximum Performance Mix Design for Each Situation

REFERENCE: Yandell, W. O. and Smith, R. B., "**Toward Maximum Performance Mix Design for Each Situation,**" *Asphalt Concrete Mix Design: Development of More Rational Approaches, ASTM STP 1041*, W. Gartner, Jr., Ed., American Society for Testing and Materials, Philadelphia, 1989, pp. 106–114.

ABSTRACT: Increasing rutting and cracking life is a financially rewarding exercise for the road authority. It appears that pavement life can be increased by paying particular attention to the plastic, as well as the elastic, behavior of asphalt concrete mixes and to their relationships with the base and subgrade of the road. The authors have selected Section 2 from the Department of Main Roads, New South Wales Rooty Hill Field Trial site near Sydney to demonstrate theoretically how both rutting and fatigue cracking life can be greatly extended by adjusting the elastic and plastic properties of the 200-mm-thick asphalt concrete pavement. Instead of formulating different mixes, the properties were varied by altering, in simulation, the temperature of the asphalt concrete. It was found that, for this particular pavement design, maximum rutting and cracking life can be achieved by grading the asphalt concrete properties from stiff nonplastic at the surface down to soft elasto-plastic at the bottom of the asphalt concrete. By contrast, the reverse grading allowed only 12 000 standard axle passes before fatigue failure. When representative field conditions were simulated, the mechano-lattice theoretical predictions were borne out by measurement.

KEY WORDS: Permanent deformation, fatigue cracking, pavement life, elasto-plastic surface, rutting, mechano-lattice analysis, computer simulation, stresses, asphalt concrete, mix design, asphalt specifications

In 1981, the Department of Main Roads, New South Wales, in cooperation with the Australian Road Research Board, constructed an experimental test pavement in Rooty Hill Road Plumpton, an outer western suburb of Sydney. The authors have been engaged in monitoring and predicting the performance of a number of sections of this road as described elsewhere [*1–6*].

One of the more interesting aspects of this task involves the prediction of cracking and rutting performance of the three asphalt concrete sections; since temperature and asphalt concrete mix data are available. In fact, the measured temperature of the asphalt concrete varies between 10 and 40°C (occasional higher temperatures could be expected) so its behavior ranges from near elastic to very plastic for the particular mix used. Also, for some of the time, negative and positive temperature gradients develop.

With available results from appropriate material tests, prediction of cracking and rutting lives of this ever-changing elasto-plastic road is relatively simple when using the mechano-lattice stress-strain analysis [*7*].

---

[1] Senior lecturer, School of Civil Engineering, University of New South Wales, Kensington NSW 2033, Australia.

[2] Senior scientific officer, Parramatta Division, Department of Main Roads, New South Wales, Parramatta NSW 2150, Australia.

The authors believe that a useful investigation of mix design could be had from this theoretical and practical study, involving the effect of temperature regimes: but instead, viewing the states of the asphalt concrete dictated by temperature pattern as being wrought by mix design. Both plasticity and rigidity at a given temperature can be determined by presetting the viscosity and quantity of the binder and by the grading and shape of the aggregate.

Thus, this study compares the theoretical effect of using mix designs for each asphalt concrete layer in particular pavements with field performance of Section 2 of the Rooty Hill test road near Sydney, Australia. The responses compared were straight-edge rutting and fatigue cracking. This paper is primarily a sensitivity analysis but some field comparisons are made.

Design of Section 2 Pavement

The thickness design of Section 2 is shown in Fig. 1. It consists of 200 mm of asphalt concrete overlying a weak imported subgrade resting on a relatively strong subgrade. It is an unusual pavement since the imported subgrade (hereafter called "base") had been deliberately chosen to be soft and plastic so that the overlying asphalt concrete will have a severe test. The high compliance and plasticity of the base have suppressed in magnitude the growth of residual stresses and strains as compared with more conventional flexible pavements.

Material Properties

The mechano-lattice analysis calculates the stress-strain behavior of multilayered elastoplastic roads in three dimensions as it is traversed in one direction by a standard wheel load. It is necessary to input both elastic and plastic parameters into the analysis. Elastoplasticity is assumed in the analysis. Table 1 shows the input values. The elastic parameters were determined by Kinder of the Australian Road Research Board using a 0.1-s stress wave width. The plastic strains for 700 000 applications of a 40 kN compressive load were determined by extrapolating results obtained from a total of 10 000 load applications by an automated repeated load universal testing machine.

Moduli at 10 and 40°C were determined using the published relationships proposed by Croney [8] and the laboratory measured moduli at 22°C (as determined by Kinder). Plastic behavior at 40°C was based on relationships between the plastic behavior at 22 and 40°C as determined by the authors in laboratory testing. Only elastic behavior was assumed to occur at 10°C. The latter was considered reasonable because of the mix stiffness and observed field behavior at low temperatures. It will be noted that plasticity increased with

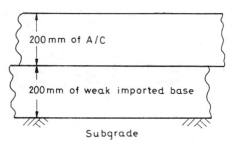

FIG. 1—*Pavement design of Section 2—Rooty Hill Test Pavement.*

TABLE 1—*Pavement material properties.*

	Asphalt Concrete			Wet Imported Subgrade	Subgrade
	10°C	22°C	40°C		
Permanent strain after 700 000 repetitions of a 50-kPa compression	0	0.0057	0.0114	0.384	0.155
Elastic modulus for 0.1-s wave widths, MPa	12 000	4488	1000	43	136
Poissons ratio	0.35	0.35	0.35	0.40	0.45

temperature. It is assumed that plasticity is proportional to the magnitude of repeated stress. This is not always true in practice.

Analysis Method

Stress Determination

The mechano-lattice multilayer analysis enables the distributions of both transient and residual stresses to be plotted. As an example, Fig. 2 shows the way the residual compressive stress at the bottom of the asphalt concrete at 22°C increases in three calculation passes. This increasing residual compression *reduces* the transient longitudinal tension in the bottom of the asphalt concrete beneath the rolling wheel.

The value of the longitudinal transient tension at the end of 700 000 standard axle passes is determined by extrapolating through the individual values of one, two, and three calculation passes [7]. Figure 3 shows that the transient longitudinal tension beneath the wheel at the bottom of the asphalt concrete has decreased from 605 to 230 kPa. Bottom fatigue cracking life has thus been increased. However, this advantage has been more than offset by the build-up of residual longitudinal tensile stresses in the top of the asphalt concrete. Figure 3 shows that its value has increased from zero before trafficking up to 1450 kPa at 700 000 standard axle passes. Thus, the number of standard axle passes to cause lateral cracks is even less than that predicted from the constant 605 kPa, if one assumed constant elasticity.

The fatigue-causing stresses, for the five cases investigated, are similarly determined and are set out in Table 2.

Fatigue Life Determinations

Fatigue life envelopes for the three elastic moduli involved with $\log_{10}$ (repeated asphaltic concrete strain) versus $\log_{10}$ (number of repetitions) are shown in Fig. 4. The envelopes were determined by Monismith and McLean [9]. It was necessary to convert the fatigue-causing stresses in Table 2 to fatiguing "strains" shown in Table 3. The conversion was made assuming linearized elasticity. The fatigue life in standard axle passes was determined at the intersection of the rising or falling strain and the fatigue envelope.

Discussion

Table 2 shows the stresses and strains in the asphalt concrete after 700 000 standard axle passes. The great variation in stress is not reflected in the equivalent strains. This is due

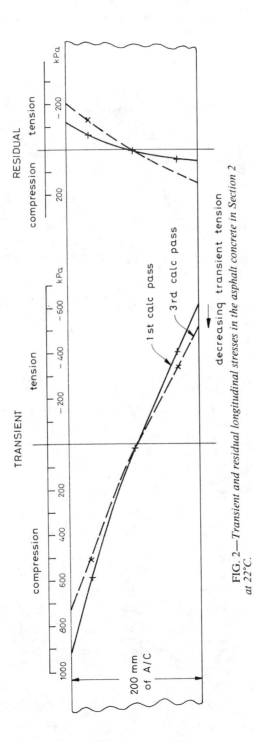

FIG. 2—*Transient and residual longitudinal stresses in the asphalt concrete in Section 2 at 22°C.*

ASPHALT CONCRETE MIX DESIGN

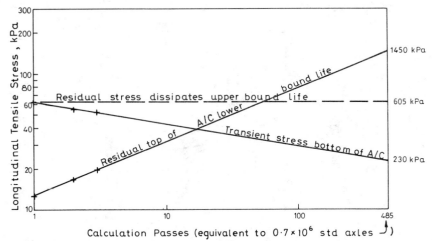

FIG. 3—*Increasing transient and decreasing residual longitudinal stresses over 700 000 standard axle passes of Section 2 at 22°C.*

to the great variation in elastic modulus (12 000 MPa down to 1000 MPa for temperatures ranging from 10 to 40°C).

Accuracy of Predictions

1.2-m Straight Edge Rutting—A temperature of 22°C is close to the weighted mean annual air temperature (w-MAAT) for that district. Predicted rutting for 22°C was 4.7 mm after 700 000 standard axle passes. Measured straight edge rutting of the order of 10 mm

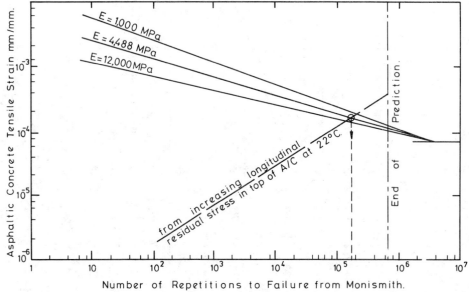

FIG. 4—*Prediction of cracking from fatigue envelopes.*

TABLE 2—*Traffic-induced residual and transient stress and strain. (Figures in brackets are for elastic case.)*

	Uniform Temperature with Asphalt Concrete Depth										
	10°C		22°C		40°C		40°C at Top Grading to 10°C at Bottom		10°C at Top Grading to 40°C at Bottom	
Type of Stress/Strain	Stress, kPa	Equivalent Strain	Stress, kPa	Equivalent Strain	Stress, kPa	Equivalent Strain	Stress, kPa	Equivalent Strain	Stress, kPa	Equivalent Strain
Residual longitudinal top of asphalt concrete	+75 (0)	0.00006 (0)	−1420 (0)	0.000316 (0)	−222 (0)	0.00022 (0)	−59 (0)	0.000005 (0)	−141 (0)	0.00001 (0)
Residual lateral top of asphalt concrete	+272 (0)	0.00002 (0)	−407 (0)	0.00009 (0)	+22 (0)	0.000021 (0)	+2.8 (0)	0.00001 (0)	+303 (0)	0.00002 compression
Transient longitudinal bottom of asphalt concrete	−803 (−727)	0.000066 (0.000060)	−230 (−605)	0.000038 (0.000133)	−386 (−421)	0.000385 (0.000420)	−1482 (−1305)	0.000123 (0.000107)	−90 (−88)	0.000089 (0.000086)
Transient lateral bottom of asphalt concrete	−1022 (−768)	0.000083 (0.000063)	−627 (−652)	0.000139 (0.000149)	−414 (−408)	0.000413 (0.000407)	−2124 (−1332)	0.000176 (0.000149)	−97 (−81)	0.000095 (0.000085)

TABLE 3—*Rutting and cracking performance. (Figures in brackets are for elastic case.)*

	Uniform Temperature with Asphalt Concrete Depth			40°C at Top Grading to 10°C at Bottom	10°C at Top Grading to 40°C at Bottom
	10°C	22°C	40°C		
1.2-m Straight-Edge Rutting at 700 000 standard axles (mm)[a]					
	1.2	4.7	11.4	2.3	2.3
Life Before Onset of Fatigue Cracking (standard axles)					
(a) Top Cracking Due to Residual Tension[b]					
Lateral cracks	∞	160 000	∞	∞	∞
Longitudinal cracks	∞	10^6	350 000	∞	∞
(b) Bottom Cracking Due to Transient Stress[b]					
Lateral cracks	∞ (∞)	10^8 (720 000)	50 000 (40 000)	500 000 (600 000)	2.4×10^6 (2.5×10^6)
Longitudinal cracks	10^6 (∞)	450 000 (300 000)	45 000 (45 000)	150 000 (700 000)	1.8×10^6 (1.8×10^6)
Quantitative Asphalt Concrete Properties					
	stiff nonplastic	moderate rigidity and plasticity	low rigidity and high plasticity	soft and plastic on top; stiff and nonplastic below	stiff and nonplastic on top; soft and plastic below

[a] The measured straight-edge rutting was of the order of 10 mm after the passage of the equivalent of 700 000 equivalent standard axles.
[b] Longitudinal cracking was observed in the field after the passage of the equivalent of 300 000 equivalent standard axles; lateral or transverse cracking developed in the field after the passage of between 560 000 and 610 000 equivalent standard axles.

was measured in the field. The discrepancy between the measured and predicted values may be due to the higher than expected moisture content in the imported base.

Fatigue Cracking—One boundary for the feasible cracking predictions is when perfect elasticity is assumed. The other boundary is yielded when residual stresses and strains are assumed not to dissipate.

If elasticity is assumed, longitudinal cracks would be expected at 300 000 standard axle passes.

If residual stresses and strains are assumed *not* to dissipate, lateral cracks due to residual longitudinal tension in the top of the asphalt concrete would be expected at only 160 000 axle passes.

A single isolated longitudinal crack was detected in the field trial section, in the inner wheelpath, after the passage of the equivalent of about 300 000 standard axles. Lateral cracking developed at between 560 000 and 610 000 passages of equivalent standard axles.

Effect of Asphalt Concrete Property Gradient

When the asphalt concrete is elasto-plastic and compliant on top, grading down to stiff and elastic at the bottom, longitudinal cracks should commence at 150 000 standard axle passes. However, when the asphalt concrete is stiff and elastic on top, grading to elasto-plastic and compliant at the bottom of the asphalt concrete, the minimum cracking life is 1.8×10^6 standard axles.

The property gradients here were achieved in simulation by assuming temperature gradients with depth. Similar effects can be achieved by modifying mix parameters: binder grade, binder content, and aggregate grading.

Lessons to Be Learned for Mix Design

For this particular pavement design with its very weak base, the best alternate strategies in choosing asphalt concrete mixes include the following.

1. Uniform Properties with Depth—The stiff, near-elastic asphalt concrete (similar to Section 2 asphalt concrete at 10°C) gives greatest cracking life at 3×10^6 standard axle passes and minimum 1.2-m straight-edge rutting.

2. Grading Properties of Asphalt Concrete with Depth—The stiff, near-elastic asphalt concrete on top, grading to soft, elasto-plastic at the bottom of the asphalt concrete gives a minimum cracking life of 1.8×10^6 standard axle passes. This compares with only 150 000 for the reverse gradient.

Conclusion

It may be worthwhile to design the elastic and plastic gradients in the asphalt concrete to maximize cracking and fatigue life. The ideal values of the elastic moduli and of the plasticity depend on the thickness design of the pavement and on the elasto-plastic properties of the underlying materials. Elasticity and plasticity can be adjusted by modifying binder grade, binder content, and by changes of the gradient and shape of the aggregate.

Acknowledgments

The authors wish to thank B. G. Fisk, Commissioner for Main Roads, NSW, for permission to publish this paper and staff of both the Department of Main Roads and Australian Road Research Board who performed the testing and conducted the surveys.

References

[1] Smith, R. B. and Yandell, W. O. *Computers and Geotechnics*, Vol. 2, 1986, pp. 23–41.
[2] Smith, R. B. and Yandell, W. O. *Journal of Transportation Engineering*, Vol. 112, 1986, pp. 649–653.
[3] Smith, R. B. and Yandell, W. O. in *Numerical Models in Geomechanics*, G. N. Pande and W. F. Impe, Eds., M. Jackson & Son, Ltd., Redruth, UK, 1986, pp. 577–586.
[4] Smith, R. B. and Yandell, W. O. in *Constitutive Laws for Engineering Materials: Theory and Applications*, C. S. Desai, E. Krempl, P. D. Kiousis, and T. Kundu, Eds., Elsevier Science Publishing Co. Inc., New York, Vol. 2, 1987, pp. 1305–1312.
[5] Smith, R. B. and Yandell, W. O., "Performance of a thin pavement, and modelling by mechano-lattice analysis," *Proceedings*, New Zealand Roading Symposium, Wellington, Vol. 2, 1987, pp. 317–323.
[6] Smith, R. B. and Yandell, W. O., "Predicted and Field Performance of a Thin Full Depth Asphalt Pavement Placed over a Weak Subgrade," *Proceedings*, 6th International Conference on the Structural Design of Asphalt Pavements, Ann Arbor, Vol. 1, 1987, pp. 443–454.
[7] Yandell, W. O., "How the Plastic Behavior of Asphalt Mixes Influences Pavement Life," *Asphalt Rheology: Relationship to Mixture, ASTM STP 941*, O. E. Briscoe, Ed., American Society for Testing and Materials, Philadelphia, 1987, pp. 76–98.
[8] Croney, D. *The Design and Performance of Road Pavements*, Her Majesty's Stationary Office, London, 1977.
[9] Monismith, C. L. and McLean, D. B., "Structural Design Considerations," *Proceedings*, Association of Asphalt Paving Technologists, Cleveland, Vol. 41, 1972, pp. 258–305.

Raymond Charles[1]

Asphalt Concrete Mix Design in the Caribbean

REFERENCES: Charles, R., "Asphalt Concrete Mix Design in the Caribbean," *Asphalt Concrete Mix Design: Development of More Rational Approaches, ASTM STP 1041*, W. Gartner, Jr., Ed., American Society for Testing and Materials, Philadelphia, 1989, pp. 115–136.

ABSTRACT: Integration of the asphalt concrete mix design procedure can be achieved through interaction between pavement distress identification, desired mix design parameters and selection of criteria, and territorial constraints including economic, technological, physical, and human resources capabilities. Surveys and investigations in Trinidad and Tobago have identified the critical types of distress as rutting and fatigue cracking, the occurrence of which are dependent on subgrade type and high pavement temperatures. A review and selection of related proven criteria from developed countries has been done to account for them. Structural analysis of full depth asphalt pavement sections has been used to generate strain-stiffness profiles for both types of distress and these profiles were combined with the selected criteria to establish a target mix stiffness. Through trial testing with a selected binder, and use of the Shell Nomograph analysis, it is shown that the target mix stiffness can be converted into stability criteria that can be substituted into Asphalt Institute criteria to provide a complete set of mix design specifications. The procedure has worked well over the past nine months.

KEY WORDS: mix design, constraints, rutting, fatigue (materials), subgrade type, strain, stiffness, thickness, binder, stability, criteria, asphalt concrete, asphalt specifications

Over the past two decades, Trinidad and Tobago had employed foreign consultants from developed territories to design approximately 200 km of major roadways [1] for which the asphalt concrete mix design parameters and criteria were handed down [2,3] as opposed to being developed locally. About 200 km of additional roadways are planned up to the year 2000. Economic and social conditions now require the entire design to be done by local consultants who need a more rational approach towards the asphalt concrete mix design.

At the beginning of 1987, the Ministry of Works, Settlements and Infrastructure, of the Trinidad and Tobago Government, requested the Civil Engineering Department of the University of the West Indies (UWI) to devise a mix design procedure that would allow for pavement distress, incorporate the obtainable level of technology in the country, and impart greater reliability to the performance of the mix.

This paper, which presents the results of the study to date, is based on investigations and surveys conducted by UWI over the past ten years, as well as a review and selection of proven methods techniques and criteria for relevant asphalt concrete technology acquired from developed territories and adapted to suit local conditions of traffic and materials.

---

[1] Lecturer, Department of Civil Engineering, University of the West Indies, St. Augustine, Trinidad.

116 ASPHALT CONCRETE MIX DESIGN

The Integrated Mix Design System

The process in choosing an asphalt concrete mix design involves the identification of distress levels in existing pavements, selection of expected critical conditions of performance, and subsequent execution of trial mix testing according to the selected test method. Our experience has shown that integration of the process as a whole requires interaction of the essential activities and input considerations or constraints. Figure 1 presents the integrated mix design system that has been developed in Trinidad and Tobago and has worked well over the past nine months in providing acceptable mixes.

The first activities begin with a visual distress survey of existing pavements of known service life in which the mixes are identified by type, that is, grading and asphalt content

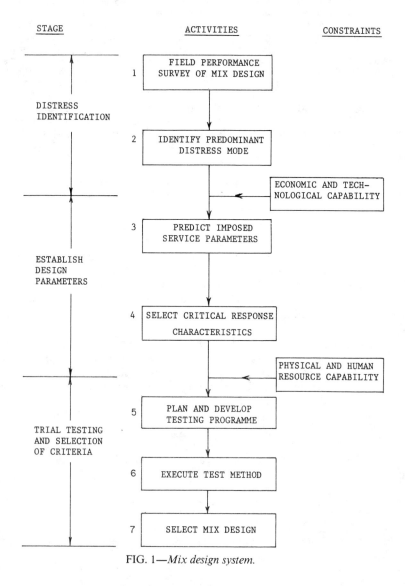

FIG. 1—*Mix design system.*

and the corresponding distresses recorded for particular terrain category according to their classification, severity, and frequency levels. The survey results are then evaluated and the predominant distress modes for the different types of mixes and terrain are identified.

Before moving on to the next stage, the economic and technological resource capability of the territory must be considered. The prime economic considerations include the project budget and the expected service life of the mix, while the technological considerations include the reliability of traffic assessment and the availability of fatigue, deflection, and prediction subsystems, and their associated precision and sophistication. In other words, do underdeveloped and developing countries have the economic and technical wherewithal equal with the state of the art in the developed countries?

In the third activity, the expected imposed service parameters are established. These include the projected traffic in terms of load category (heavy, medium, or light axle loads) and the number of equivalent standard axles (ESA) and associated percentage overload, all within the considered life of the pavement. Consideration is also given to mean pavement temperatures that would account for daily and seasonal climate variations.

The fourth activity concerns the generation of the desired response, that is, characteristics that are ultimately embraced in the materials specification and the mix design criteria and are geared to resist the predominant distress mode.

The physical or laboratory resource capability restricts the selection and execution of the test method to be adopted for trial mix evaluation. This consideration is of crucial importance because the stiffness modulus obtainable from testing machines and equipment in the territory must be of the same character as those used in the stress-strain computations. In addition, the laboratory personnel must be capable of producing reliable results, that is, they must be well trained and experienced in the use of the machines and equipment. This combination of equipment and personnel is difficult and, in some cases, impossible to locate in most Caribbean territories for state-of-the-art test methods and techniques [4]. In Trinidad and Tobago, the Marshall Test Method [5] and the Shell Nomograph stiffness determination [6] are used in asphalt concrete mix evaluation because the test procedure for the Marshall method is relatively straight-forward and simple and the input parameters for the Shell stiffness determination are easy to generate.

The fifth activity is the planning and development of a testing program to fit the standards of the suitable test method and the in-service mix conditions where possible. The program details the number and variety of trials mixes, the temperatures and conditions of loading with mix types, and the temperature variations.

The program is executed as the sixth activity, and the ideal or design mix is selected according to the set mix design criteria to complete the seventh activity.

Distress Identification

Modes

Flexible pavement distress has been categorized into load associated (LA) and non-load associated (NLA) distresses by Monismith and McLean [7]. It is believed that the relevant category for consideration is the LA distresses because the NLA distresses are usually caused by subgrade and environmental factors that are not essentially due to the properties of the surfacing mix. The LA distresses in Trinidad and Tobago are classified as non-structural and structural.

Non-structural distresses are considered as distresses of the first order that usually reflect the integrity of the mix and are caused by inadequate selection and improper proportioning of aggregate and asphalt components. They are characterized by poor compactibility, progressive loss of aggregate and mix, shrinkage, and bleeding, which are respectively manifested as raveling, potholes, shrinkage cracks, and poor skid resistance.

Structural distresses are essentially second-order distresses that reflect the composite effect of aggregate gradation, asphalt content, and level of compaction, and are caused by the inability of a derived property or response mechanism to safely meet the imposed loading. They are normally characterized by inadequate mix stiffness, and the types of distress manifested are rutting (excessive deflection) and fatigue cracking (failure in tension).

Surveys

Distresses are noted from subjective and limited physical observations along sections of existing major routes that span a variety of terrain categories in the islands. Survey techniques employed include manual traffic counting, visual observation, distance measurement by taping, defining proportion of distress in the route section, rut depth measurement, coring of the surface for mix type identification, and ultrasonic pulse velocity (USV) measurement with a portable ultrasonic non-destructive digital indicating tester (PUNDIT). The USV measurement has been adopted as an overall condition indicator and the procedure is nondestructive. The equipment is relatively cheap, lightweight, and simple to operate, and the obtained USV may be converted to a dynamic stiffness modulus, E, according to the equipment manufacturers [8] given the unit weight and an assumed Poisson's ratio for the mix. It also facilitates the determination of layer thickness and discrimination between surface and fatigue, or deep, cracks. Signal frequencies of 20 and 54 kHz are used with a pulse frequency of 10 Hz according to the physical procedure outlines in the instrument manual [8].

Findings

The terrain in Trinidad and Tobago ranges from terraces, plains, peneplains, basins, and swamps. Approximately 70% of the country is covered by clays. In most cases, the natural soils are used as subgrades, and the types include high to low-plasticity clays, sandy clayey gravels, coheshionless sands, and earthfill that are used in embankments for routes in swamp terrain.

Continuous surveys and enquiries over the past ten years have shown the following.

1. The mixes used in Trinidad and Tobago are fairly standard in grading, and generally fit the AASHO Test Road bindercourse and surface course grading specifications [9].
2. Nonstructural distress, quality control, and materials selection are directly related; route sections that were laid with poor or no quality control manifest the spectrum of this type of distress. The surface characteristics and skid resistance of the pavement are greatly influenced by the type of aggregate and binder used. With standard asphalt cement (AC) binders, crushed limestone aggregates exhibit much lower and unfavorable skid resistance than do natural aggregates [10] primarily because of the higher polishing effect of traffic on the limestone particles. Compensation is obtained in some cases by using Trinidad Lake Asphalt (TLA) binder, which has substantial inherent mineral matter that imparts a rough surface texture to the exposed bitumen in the worn matrix [11]. On the other hand, mixes with natural aggregates tend to exhibit a greater amount of raveling than those with crushed limestone aggregates.
3. The frequency and occurrence of structural distress appears to be dependent on terrain and subgrade type. On the average, sections located in basins and plains with clay subgrades exhibit higher rut depths than do those located on cohesionless

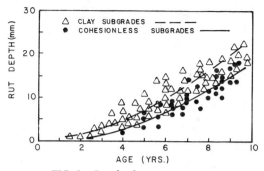

FIG. 2—*Rut depth versus pavement age.*

subgrades (Fig. 2) and can be as high as 22 mm for eight to nine-year-old pavements. Sections located on terraces and embankments with cohesionless subgrades exhibit a higher degree of percentage fatigue cracking in the wheel tracks (Fig. 3). Both observations refer to the same pavement sections.

4. Once one type of structural distress occurs, the other type follows; that is, rutting precedes cracking in clay subgrades and vice versa in cohesionless subgrades (Figs. 2 and 3).
5. The occurrence of unexpected commercial traffic accelerates distress manifestation as in the case of the Uriah Butler Highway North Bound Lane, Section 2, which is the major fast link between the recently opened Point Lisas Port in Central Trinidad and the north of the island. The sudden introduction of additional truck traffic has brought about a significant increase in structural distress types within eight months to one year. Fatigue cracking in the wheel track increased by 40 percentage points and average rut depths increased from 13 to 18 mm. The average daily ESA has increased by approximately 50%. This pavement has had eleven years prior service with one 10-cm overlay in its sixth year.
6. Based on three years of USV pavement-surface profile surveys (totaling 16 to date), an interim condition classification has been developed correlating USV, distress level (rut depth and percentage cracking), and visual condition rating (see Table 1). The

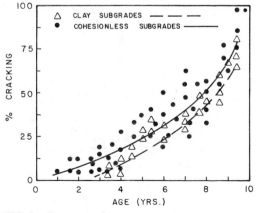

FIG. 3—*Percent cracking versus pavement age.*

120   ASPHALT CONCRETE MIX DESIGN

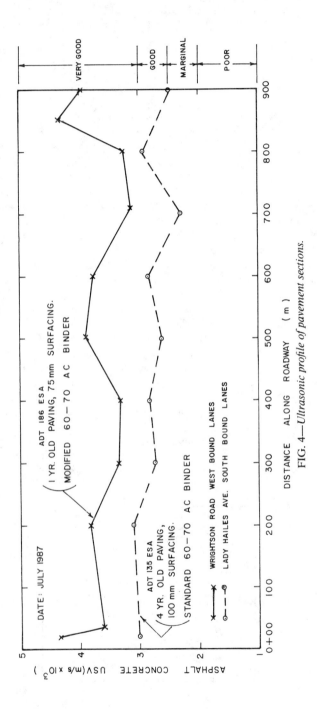

FIG. 4—*Ultrasonic profile of pavement sections.*

TABLE 1—*Pavement mix USV condition classification.*

	Distress Level		
USV, m/s	Average Rut Depth, mm	Average Fatigue Cracking, %	Condition Rating, visual
<2000	15	50 to 100	bad
2000 to 2500	10 to 15	25 to 50	marginal
2500 to 3000	5 to 10	0 to 75	good
>3000	0 to 5	0	very good

classification refers to different levels of either distress type that may occur for a range of USV measurements. The results of two USV profile surveys, along two of the pavement sections used in the development of the classification system, are presented in Fig. 4. The ultimate aim is to be able to identify a mix stiffness level from visual condition inspection.

7. Derived stiffness moduli, E, obtained from USV measurements show significant variation with daily temperature fluctuations (Fig. 5). The E values vary from 9.5×10^9 N/m² at 27.5°C (81.5°F) to 5×10^8 N/m² at 35°C (95.0°F). The net effect appears to be a drastic softening of the mix with increasing temperatures.
8. All together, structural distress predominates over nonstructural distress.

Imposed Service Parameters

Traffic Composition

In Trinidad and Tobago, the actual spectrum of commercial traffic is usually established from two types of counts: (1) automatic counting to define the 16 and 24-h traffic volume, (the ADT) and (2) manual 16-h counts to establish the commercial traffic as an overall percentage of ADT. In the manual counts, the types of commercial traffic or trucks are usually recorded according to load classification (light, medium, or heavy) and, in a few cases, the size of payload is noted as empty, half, or fully loaded.

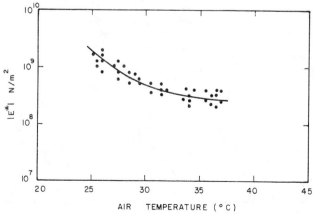

FIG. 5—*Effect of air temperature on pavement stiffness modulus.*

TABLE 2—*Commercial traffic composition.*

Load Class	Unit Type	Truck Traffic Composition, %	ADT
Light—two axles single tires	rigid	25.5	5.00 ± 20%
Medium—two axles dual rear tires	rigid	64.75	14.20 ± 40%
Heavy—three axles or more	rigid articulated	9.75	2.00 ± 5%

The light trucks are normally pick-up trucks and vans, the medium trucks are typical freighters and dumpers, and the heavy trucks are the fuel tankers, haulers, and containerized cargo units. Table 2 presents the average results to date of a six-year survey of the commercial traffic spectrum, excluding payload, for Trinidad and Tobago's major road network. The medium and heavy trucks account for 75% of the commercial traffic, and excessive pavement damage is usually due to these two categories.

Equivalent Standard Axles

Pavement damage in the Caribbean appears to be a direct result of inadequate vehicle load regulations and attendant control, with consequent excessive ESA values and overloaded trucks. In Trinidad and Tobago, existing vehicle load regulations [12] impose a 15-ton maximum gross weight (MGW) per vehicle with an 8-ton maximum single axle load specification. Enquiries for Barbados and Jamaica reveal that the respective MGW values are 14 and 21 tons, respectively, and, there are no specifications for tire width that affects the transmitted pressure. In the entire Caribbean, vehicle load control surveys are never done. In Trinidad and Tobago, damage is evaluated through an average ESA per commercial vehicle that is estimated from the manual counts with the assumption that the load per vehicle is distributed in a 66:34 rear to front axle proportion. A probable overall load percentage is sometimes included in this second-hand type ESA analysis.

This study has shown so far that reliable ESA values can only be obtained from axle load

TABLE 3—*Summary of truck axle load survey.*

Description	Medium, two axles	Heavy, three axles	Heavy, four axles
Total trucks	954	82	127
Trucks sampled	167	12	30
% sampled	17.5	14.2	23.6
Number of commercial axles	334	36	129
Number of ESA[a]	75	37	95
Axle damage factor[b]	0.2	1.0	0.7
Vehicle damage factor[c]	0.5	3.1	3.1
% vehicles with axles > 8000 kg	12.0	33.2	21.7
% axle > 8000 kg	6.3	22.2	21.7

[a] ESA = (axle weight ÷ 8160 kg)^4.
[b] Number of ESA ÷ number of commercial axles.
[c] Number of ESA ÷ number of vehicles.

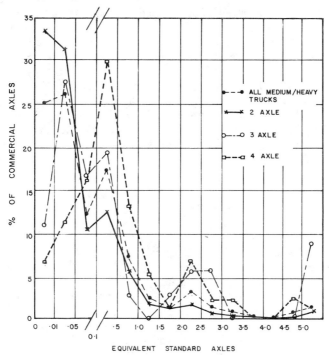

FIG. 6—*Distribution of equivalent standard axles.*

surveys that would then define the extent and degree of overloading. The results of a typical axle load survey and ESA distribution analysis for medium and heavy commercial vehicles on a major route in Trinidad and Tobago [*1*] are presented in Table 3 and Fig. 6. The data in Table 3 show that

(*a*) vehicle damage factors vary between 0.5 and 3.1 for medium and heavy vehicles; and
(*b*) there is significant overloading in these vehicle classifications, between 12.0% for medium trucks to as high as 50.0% for heavy trucks.

Clearly, an average ESA or vehicle damage factor per commercial vehicle and an assumed overall percentage overload would produce severe errors. The prime reasons for such extensive overloading are excessive payload (not improper distribution) and the cost of the imported trucks, because most truckers want to maximize trip financial returns to meet the high vehicle installment payments.

Further damage definition by the ESA distribution given in Fig. 6 affords more accurate evaluation of the total truck damage, which can then be projected over the design life of the pavement.

Projected ESA

Estimating the amount of traffic at some point in the future is normally done by means of one of several methods [*13*] each depending essentially on the traffic growth rate, i, the period under concern, n, and the initial ESA.

It has been common practice in Trinidad and Tobago to consider the growth rate as a percentage annual increase obtained from the year in which the design is done (the base year), as opposed to an average annual growth rate obtained from a fixed base period or previous base year. This can lead to severe error as the growth rate depends upon several factors such as commodity growth, gross domestic product (GDP), and the effect of national or governmental economic policies and plans. This study has examined this problem with respect to the base year and the GDP, for major commercial routes in Trinidad and Tobago and the results to date are presented in Figs. 7 and 8.

Figure 7 shows the behavior of the average annual growth factor calculated from different base years over a 17-year period. Figure 8 shows how the dependence of the growth rate on the GDP, expressed as

$$\text{elasticity} \left[\frac{\text{traffic growth rate}}{\text{GDP growth rate}} \right]$$

behaved over a 15-year period. The increases in growth rate (Fig. 8) between 1976 and 1982 reflect the period of economic boom in Trinidad and Tobago, while the drop off beyond 1982 characterizes the subsequent economic recession. Also noticeable is the dependence of the growth rate gradient on the length of the analysis period. However, they all appear to taper off to a common steady gradient some years after. Clearly, the longer the base period, the more steady and more reliable is the growth rate. Such steadiness is reflected by the behavior of the elasticity over a similar period. Therefore, it has been proposed that designers use the average annual growth rate calculated from a base period of no less than ten years prior to the year of the design, for more reliable projections.

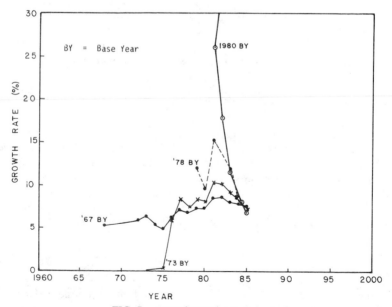

FIG. 7—*Annual growth rate behavior.*

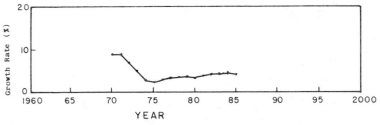

FIG. 8—*Elasticity behavior.*

Temperature

Trinidad and Tobago has a bi-annual climatic cycle, that is, a wet and a dry season, each lasting for approximately six months. Average maximum daily temperatures (6 a.m. to 6 p.m.) show almost no variation from season to season [*14*] and center around 30.9°C (87.6°F) with a standard deviation of 0.36. Minimum daily temperatures that occur at night (6 p.m. to 6 a.m.) average 22.42°C (72.3°F) with a standard deviation of 0.47. Consequently, the resulting variation in pavement temperature, characterized by surface temperature in this study, would also remain relatively small throughout the year.

A simple analysis of 141 field temperature measurements was done according to the method outlined by Witczak [*15*] to establish a relationship between the measured mean air temperature (MAT) and the mean pavement temperature (MPT). The simple linear empirical correlation obtained is of the following form

$$\text{MPT} = 1.07 \text{ MAT} \pm t_{\alpha,140} 3.53 \tag{1}$$

where

MPT = pavement temperature, °C;
MAT = air temperature, °C;
α = probability level;
$t_{\alpha,140}$ = t value for pavement temperature masurements; and
3.53 = standard deviation of MPT.

The conservative upper limit value of MPT was selected for use and for a probability of 0.80

$$\text{MPT} = 1.07 \text{ MAT} + 4.53 \tag{2}$$

for a measured maximum mean daily air temperature of 31.2°C (88.2°F) where the mean pavement temperature is 37.9°C (100.2°F). Use of the average maximum here is considered to provide some added safety.

Response Characteristics

The essential material response characteristics needed to resist the predominant structural and non-structural LA distresses due to the imposed service parameters are mix stiffness, stability, and the type of materials used in the mix, that is, aggregate gradation, physical, and mechanical properties, and the grade of binder. The procedure employed in generating the characteristics is schematically outlined in Fig. 9 and is based essentially on

126 ASPHALT CONCRETE MIX DESIGN

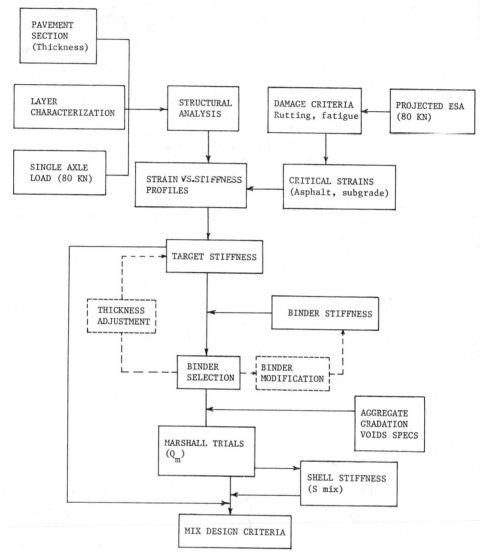

FIG. 9—*Schematic outline of response characteristics determination.*

stiffness selection and a materials selection, and can, if desired, incorporate a full-depth asphalt thickness selection.

The required mix stiffness involves the integration of a structural analysis done to generate respective strain versus stiffness profiles for rutting and fatigue cracking, and damage characterization that involves selecting appropriate rutting and fatigue cracking criteria from which the average permanent or critical strains would be obtained for the projected ESA. These strains are then converted into target stiffnesses through the computed strain-stiffness profiles. Aggregate and binder selection then follow, and these allow for final specification of the desired stiffness level.

Structural Analysis

Despite the current availability of a variety of computer programs for the structural analysis [16], the best that can be done at present is to use the layered systems program, ELSYM 5 [17], which has the following attributes.

1. It can accommodate multiple layers to a maximum of five.
2. It includes provision for superposition of dual-tire loads to a maximum of five.

The inputs to be used are given in Fig. 10, which shows a two-layer pavement section, layer parameters, and a single-axle dual-tire loading. A Poisson's ration of 0.35 is used for both layers, and the varied representative surface mix stiffnesses and thicknesses are used to generate a spectrum of stress-strain profiles. Typical calculated profiles for rutting and fatigue are shown in Figs. 11 and 12, respectively.

Rutting

The rutting criterion selected, which was suggested by Shell Researchers [18], is that by controlling the magnitude of compressive strain at the surface of the subgrade the permanent deformation is also controlled.

The basic theory assumes that the logarithmic relationship between the number of strain repetitions and permanent strain at a particular depth is essentially linear over a range of strain repetitions and can be described by the following equation

$$\mathcal{E}_c = IN^S \tag{3}$$

where

$\mathcal{E}_c$ = accumulated permanent strain at a particular depth,
I = intercept with permanent strain axis (arithmetic strain value),
N = number of strain repetitions, and
S = slope of the linear portion of the logarithmic relationship.

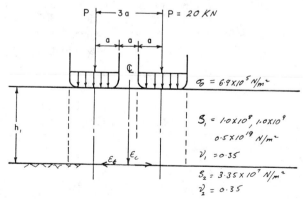

FIG. 10—*Pavement section inputs.*

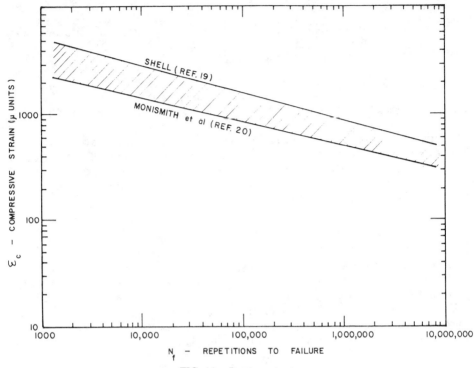

FIG. 11—*Rutting criteria.*

One average relationship relating the permissible compressive strain in the subgrade and anticipated traffic found by Edwards and Valkering [19] is

$$\mathcal{E}_c = 2.8 \times 10^{-2} \times N^{-0.25} \tag{4}$$

where $\mathcal{E}_c$ = permissible compressive strain in subgrade.

The relationship allows for dual-wheel loads and the influence of lateral wheel distribution, and is independent of axle load and subgrade modulus. The associated rut depth is of the order of 20 mm. Another form of this criterion by Shook et al. [20] that has been incorporated by the Asphalt Institute in their design procedure is

$$\mathcal{E}_c = 1.05 \times 10^{-2} N^{-0.223} \tag{5}$$

This equation is associated with rut depths up to 12.5 mm. Use of the two criteria provides a range of critical subgrade strains (Fig. 11). The Asphalt Institute criterion is the more conservative of the two; at present the Shell criterion is being used.

Fatigue Cracking

The "accepted" fatigue criterion [18] is that control of the horizontal tensile strain at the bottom of the asphalt layer, controls the incidence of cracking in that layer. Fatigue criteria are normally based on the results of three different methods [16]. They are results based

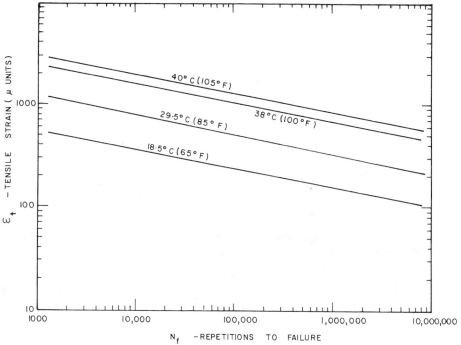

FIG. 12—*Fatigue criteria (after Ref 15)*.

on the laboratory test, results based upon an examination of existing design curves, and results based upon an elastic layered analysis of field test section performance. In all approaches, it is assumed that a linear relationship exists between the logarithm of initial strain or controlled strain and the logarithm of strain repetitions or fatigue life. The relationship has the following form

$$N_f = K\left[\frac{1}{\mathcal{E}_t}\right]^n \qquad (6)$$

where

N_f = number of strain repetitions,
$\mathcal{E}_t$ = tensile strain,
n = slope of the logarithmic relationship between fatigue life and tensile strain, and
K = antilog of the intercept of the logarithmic relationship between fatigue life and tensile strain.

Two approaches, one based on field data and the other on laboratory results, were considered. A modified form of the Kingham criteria [*15*] was selected to avoid assumptions associated with laboratory fatigue analysis such as type of test, stress conditions, temperature effects, conditions of loading rate, and time between load applications. The criteria, which is applicable to full-depth asphalt pavements, is of the form

$$N_{f(q)} = ab^{qd_1}\left[\frac{1}{\mathcal{E}_t}\right]^c \qquad (7)$$

where

$a = 1.86351 \times 10^{-17}$,
$b = 1.01996$,
$c = 4.995$,
$d_1 = 1.45$,
q = pavement temperature (°F), and
$\mathcal{E}_t$ = asphalt mix tensile strain.

Figure 12 represents the family of strain-temperature curves established from the equation.

Mix Stiffness

Typical stiffnesses and corresponding thicknesses can be obtained for rutting and fatigue cracking at mean pavement temperature by conversion of the respective critical strains through the computed strain-stiffness profiles (Figs. 13 and 14), and can be used to prepare respective thickness-stiffness relationships (Fig. 15). The selection of target stiffnesses, S_t, would then depend on the predominant distress mode, that is, whether cracking precedes rutting or vice versa. This is obtained from the results of performance surveys that gener-

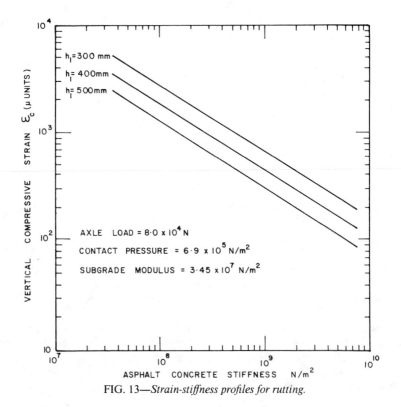

FIG. 13—*Strain-stiffness profiles for rutting.*

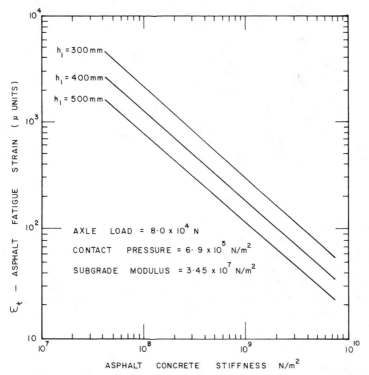

FIG. 14—*Strain-stiffness profiles for fatigue.*

ally, to date, are controlled by the terrain and the nature of the subgrade soils. Apart from the country having more clays than sands and gravels, rutting tends to present a greater driving hazard than cracking failure due to stored water in the ruts during rainfall. Consequently, rutting is presently considered as the critical distress in Trinidad and Tobago for mix stiffness selection.

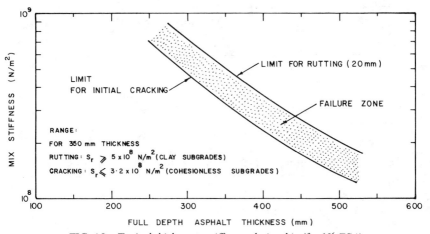

FIG. 15—*Typical thickness—stiffness relationship (for 10^6 ESA).*

TABLE 4—*Typical Trinidad and Tobago asphalt cements.*

Penetration Grade	Average[a] Penetration at 25°C, mm × 10^{-1}	Softening[a] Point (R & B), °C	Penetration Index, PI	Binder Stiffness,[b] N/m²	Mix Stiffness, N/m²
40 to 50	45	53	−0.72	5.0×10^5	2.169×10^9
60 to 70	63	47.5	−0.67	1.90×10^5	1.18×10^9
95 to 100	87	46.5	−0.74	1.35×10^5	9.3×10^8

[a] Based on three year production period (Ref *21*).
[b] Frequency 2 Hz; temperature = 38°C (100.4°F).

Component Materials

The aggregates selected for the mix normally have the general characteristics, physical properties, and grading for coarse and fine aggregate conforming to ASTM Specification for Coarse Aggregate for Bituminous Paving Mixtures (D 692-79) and ASTM Specification for Fine Aggregate for Bituminous Paving Mixtures (D 1073-81).

The binder selection is important here because it is well established [*18*] that the binder stiffness controls the mix stiffness, and for a fixed aggregate grading and level of compaction, it becomes the prime variable.

Three penetration grades of standard straight-run AC binders, 40-5-, 60-70, and 85-100 are produced in Trinidad and Tobago by Trintoc Ltd. refinery in conformance with ASTM Specification for Penetration-Graded Asphalt Cement for Use in Pavement Construction (D 946-82). A three-year production average [*18*] of penetrations, ring and ball (R&B) softening points, and resulting penetration indices, as well as stiffnesses determined by the Shell Nomograph procedure [*6*] at the mean pavement temperature of 38°C (100.4°F), are presented in Table 4 for the three grades of binders. The derived maximum probable stiffnesses (MPS) for the three binders, employing an aggregate volume concentration, C_v, of 0.9 and air voids of 3% are also given in Table 4. The respective MPS value

TABLE 5—*Consistency of a 60-70 penetration asphalt cement modified with TLA.*

Property	TLA (in 60-70 Binder), %[d]				
	0	15	30	45	100%
Penetration, 25°C, mm × 10^{-1}	69[a] (65)[b]	63 (36)	55 (26)	30 (17)	3 ...
Softening point (R&B), °C	49.7 (52.9)	51.2 (58.5)	52.7 (60.4)	54.0 (64.0)	95.3 ...
Penetration index	−0.72 (+0.16)	−0.51 (−0.03)	−0.47 (−0.33)	−1.86 (−0.46)	+1.00 ...
Stiffness,[c] N/m²	2.5×10^5 (3.5×10^5)	3.0×10^5 (1.0×10^6)	3.7×10^5 (1.7×10^6)	9.0×10^5 (4.0×10^6)	...

[a] Before mixing 69.
[b] After mixing (65).
[c] Frequency of 2 Hz; temperature-38°C (100.4°F).
[d] After Ref *22*.

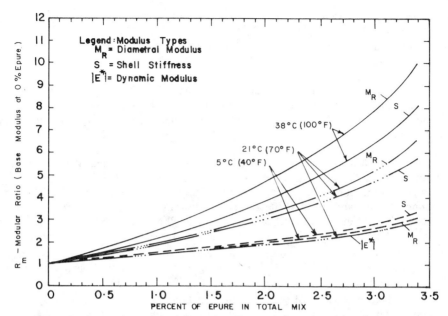

FIG. 16—*Comparison of modular ratio—Epure quality relationships between various methods of obtaining modulus (after Ref 22).*

that is closest to the respective limit of stiffness for a particular thickness directs the selection of the appropriate grade of binder. If the target stiffness is greater than that obtainable from the standard AC binders, consideration is then given to either increasing the thickness of the asphalt layer and effectively reducing the stiffness, or modifying the binder.

In Trinidad and Tobago, minor percentages of TLA are used as the modifier. Table 5 gives the basic properties of TLA, along with the resulting and derived changes in relevant binder properties for addition of 15, 30, and 45% TLA to a standard 60–70 AC [22]. The input for the stiffnesses was obtained for extracted binder after laboratory mixing with aggregate. The drastic changes in penetration and R & B softening point is attributed to the presence of TLA; this property has beneficial effects in that subsequent stiffnesses [22] are significantly improved (Fig. 16). Based on the relationship shown in Fig. 13, the desired increase in stiffness is obtained by selecting the appropriate percentage of TLA to be added to the binder.

Mix Design Criteria

The recommended mix design criteria by the Asphalt Institute [23] for parameters such as the number of compaction blows, air voids content, and voids in the mineral aggregate have been adopted by the Ministry of Works in Trinidad and Tobago since 1976 for the national road building program and have provided well-integrated mixes to this date. However, the criteria for stability and flow are always in need of adjustment to suit the anticipated traffic conditions.

The Marshall method for stability determination has been adopted in Trinidad and Tobago because:

(a) it produces satisfactory results,
(b) the apparatus is affordable and portable, and
(c) the apparatus could be adapted to existing California bearing ratio (CBR) test equipment.

In developing the stability and flow criteria, the aim is to seek agreement between the target stiffness, S_r, and the mix stiffness, S_{mix}, that are then converted to the Marshall stability or the Marshall quotient, Q_m (ratio of stability in kilonewtons to flow in millimetres), obtained from trial testing. The procedure adopted is as follows.

1. Do the Marshall test, using 50 or 75 blows per face of specimen, with the selected materials and determine Q_m for maximum stability at 60°C (140°F).
2. Repeat a minimum of ten to twelve Marshall tests on samples prepared to yield maximum stability and calculate the average or mean mix quotient, $\overline{Q}_m$.
3. Using the extracted binder properties and the Shell Nomograph analysis, determine the mix stiffness, S_{mix}, at 38°C (100.4°F) and the average or mean mix stiffness, $\overline{S}_{mix}$.
4. If S_r falls within the 95% confidence interval of $\overline{S}_{mix}$, then there is agreement and the mix is considered acceptable. If there is no agreement, then retrials using altered gradation are done.
5. The specification limits of Q_m can then be obtained from the equivalent $\overline{Q}_m$ confidence interval. The lower limit of the interval is normally selected.

This equivalence approach to relate stiffness and stability was adopted because an overall empirical correlation between Q_m and S_{mix} so far proved to be inadequate. A total of 41 Marshall tests using 75 blow compactions were done on a mix of fixed aggregate grading and a binder content of 6% using rounded-cubical and platy or angular-shaped aggregates

TABLE 6—*Typical equivalence between mix quotient and stiffness.*

	Mix A[a]		Mix B[b]		Mix C[c]	
	Q, KN/mm	S_{mix}, N/m² × 10⁹	Q, KN/mm	S_{mix}, N/m² × 10⁹	Q, KN/mm	S_{mix}, N/m² × 10⁹
Number of samples	18	18	13	13	10	10
Mean	4.27	6.37	4.42	7.93	3.77	4.97
Variability, CV%	9.47	26.94	16.65	36.64	9.0	21.2
CORRELATION, r						
	0.15		0.09		0.58	
95% confidence interval of mean	4.1 to 4.5	5.5 to 7.2	4.0 to 4.9	6.2 to 9.7	3.5 to 4.0	4.2 to 5.7

[a] Rounded–cubical aggregate and blended TLA 60–70 binder.
[b] Platy aggregate and blended TLA 60–70 binder.
[c] Rounded–cubical aggregate and 60–70 refinery bitumen.

separately. The results of the equivalence between Q_m and S_{mix} are summarized in Table 6, and while they show acceptable variability (CV) in Q_m (9.0 to 16.7%), the CV in S_{mix} is relatively high (21.2 to 36.64%) and just over twice that for Q_m in all mixtures. For mixes having the same binder but different aggregate shape, Q_m and S_{mix} do not correlate as shown by the very low correlation coefficients (0.15 and 0.09) for mixes A and B, respectively. Mix C exhibits moderate correlation (0.58) between Q_m and S_{mix}. The overall correlation between Q_m and S_{mix} for the 41 samples is 0.18, which indicates no correlation. Clearly, the empirical correlations between Q_m and S_{mix} will not be adequate. Ongoing efforts are centered around verification for 50-blow compaction.

Examination of the mean values, $\overline{Q}_m$ and $\overline{S}_{mix}$, and their 95% confidence intervals for Mixes A and B shows that while there is significant agreement between the respective $\overline{Q}_m$ and $\overline{S}_{mix}$ values, there is significant difference between the $\overline{S}_{mix}$ values, with Mix B made from platy aggregate exhibiting the higher value. Corresponding values for Mix C are much lower than $\overline{Q}_m$ and $\overline{S}_{mix}$ for Mixes A and B primarily because of the refinery bitumen binder used (see sections on Mix Stiffness and Component Materials). When this difference in values for Mixes A and B is considered together with the higher CV exhibited by Mix B, it becomes evident that a change in aggregate shape from rounded-cubical to platy can result in higher mean stiffness values with increased variability. The primary accountable factor here appears to be a higher varied particle orientation obtained with platy or angular aggregate during mixing and compaction. It is considered necessary therefore to specify aggregate shape when using the approach just outlined.

Conclusions

To date, the results of this study have shown the following.

1. Integration of the mix design process requires interaction between the essential activities and the territorial capabilities in the areas of economy technology and physical and human resources.
2. The mean daily air temperature (MAT) affects the stiffness of the pavement surface, and its relationship to mean pavement surface temperature (MPT) can be represented by

$$MPT = 1.07\,MAT + 4.53$$

for a probability of 0.80. In Trinidad and Tobago, MPT is 38°C (100°F) for a measured MAT of 31.2°C (88.2°F).
3. It may be possible to establish an overall pavement mix condition rating from ultrasonic velocity surveys, and ultimately to appraise the in situ mix stiffness levels from visual surveys if not from ultrasonic surveys.
4. Structural distress predominates over non-structural distress. Rutting precedes fatigue cracking in pavements with clay subgrades whereas fatigue cracking precedes rutting in pavements with cohesionless subgrades. As one type of distress occurs, the other follows.
5. A target stiffness range for mix design can be established to account for both rutting and fatigue cracking, from structural analyses and selected distress criteria.
6. Based on available asphalt cement binders, a maximum probable mix stiffness can be determined for a fixed aggregate grading and compared with the target stiffness range for binder selection.

7. For any one shape of aggregate, the target stiffness may be readily converted to equivalent mix stability criteria by the Marshall method and the Shell Nomograph analysis of compacted trial specimens prepared with the selected binder.

References

[1] "National Transportation Policy Study," Final Report of Republic of Trinidad and Tobago, Vol. 2A, Lea-Pal Ltd., Port of Spain, 1985, Chapter 11.5.
[2] "Contract, Specification and Bills of Quantities for the Construction of the Mucurapo Foreshore Freeway," Highways Division, Ministry of Works, Government of Trinidad and Tobago, Trintoplan Consultants Ltd. and Lee Young and Partners Ltd., 1978.
[3] "Contract and Specifications for Improvements to Princess Margaret Highway," Highways Division, Ministry of Works, Government of Trinidad and Tobago, Freeman Fox & Partners, and T.E.R. Ltd., C.E.P. Ltd., 1978.
[4] Kennedy, T. W., Adedmila, A. S., and Haas, S., "Materials Characterisation for Asphalt Pavement Structural Design Systems," *Proceedings*, 4th International Conference on Structural Design of Asphalt Pavements, Ann Arbor., MI, 1977.
[5] "Investigation of the Design and Control of Asphalt Paving Mixtures," Technical Manual No. 3-254, U.S. Army Engineer Waterways Experiment Station, Vicksburg, MS, May 1948.
[6] Huekelom, W., "Bitumen Test Data Chart for Showing the Effect of Temperature on the Mechanical Behaviour of Asphaltic Bitumens," *Journal of the Institute of Petroleum*, Vol. 55 No. 546, 1969.
[7] Monismith, C. L. and McLean, D. B., "Design Considerations for Asphalt Pavements," Report No. TE71-8, University of California, Berkeley, Dec. 1971.
[8] "PUNDIT Manual for Use with the Portable Ultrasonic Non-Destructive Digital Indicating Tester Mark IV," C.N.S. Electronics Ltd., London, 1981.
[9] Oglesby, C. H. and Hewes, L. I., *Highway Engineering*, 2nd ed., Wiley, New York, 1963, p. 596.
[10] Lyn, M. A., "A Study of the Problem of Skid Resistance in Trinidad and Tobago," undergraduate report, University of the West Indies, Faculty of Engineering, 1977, unpublished.
[11] Charles, R. in *Proceedings*, VIII World Congress on Trinidad Lake Asphalt, Port of Spain, Trinidad and Tobago, 1985, p. 6.
[12] "Motor Vehicles and Road Traffic Act," The Laws of Trinidad and Tobago, Part V, Section 28 (e) and (h), Government of Trinidad and Tobago, The Ministry of Legal Affairs, Port of Spain Trinidad and Tobago, 1978.
[13] Yoder, E. S. and Witczak, M. W., *Principles of Pavement Design*, 2nd ed., Wiley, New York 1975, pp. 171–173.
[14] "Annual Statistical Digest," Central Statistical Office, Annual Report No. 33, Ministry of Finance and Planning, Government of Trinidad and Tobago, 1986, p. 5.
[15] Witczak, M. W., "Design of Full-Depth Airfield Pavements," *Proceedings*, Third International Conference on the Structural Design of Asphalt Pavements, Ann-Arbor, MI, 1972, pp. 552–55.
[16] Monismith, C. L. and Finn, F. N. "Moderators' Summary Report of Papers Prepared for Discussion at Session III," *Proceedings*, Third International Conference on the Structural Design of Asphalt Concrete Pavements, Ann-Arbor, MI, 1972, pp. 144–170.
[17] Ahlborn, G., "ELSYM 5, Computer Program for Determining Stresses and Deformation in Five Layer Elastic System," University of California, Berkeley, 1972.
[18] Claessen, A. I. M., Edwards, J. M., Sommer, P., and Uge, P. in *Proceedings*, Fourth International Conference on the Structural Design of Asphalt Concrete Pavements, Ann-Arbor, MI, 1977, p. 41.
[19] Edwards, J. M. and Valkering, C. P., "Structural Design of Asphalt Pavements for Road Vehicles—The Influence of High Temperatures," *Highways and Road Construction*, Vol. 42, Pa 1770, Feb. 1974.
[20] Shook, J. F., Finn, F. N., Witczak, M. W., and Monismith, C. L., "Thickness Design of Asphalt Pavements," *Proceedings*, Fifth International Conference on the Structural Design of Asphalt Concrete Pavements, The Netherlands, Aug. 1982.
[21] "Quarterly Quality Control Summaries," Trintoc Ltd., Ministry of Energy, Government of Trinidad and Tobago, 1986.
[22] Witczak, M. W., "The Influence of Trinidad Epure upon the Elastic Module Behaviour Asphaltic Mixes," Consulting Report, Jan. 1978, unpublished.
[23] "Model Construction Specifications for Asphalt Concrete and Other Plant-Mix Types," Specification Series No. 1 (SS-1) 5th ed., The Asphalt Institute, Nov. 1975.